Environmental Adaptation and Eco-cultural Habitats

In this challenging and highly original book, the author tackles the dynamic relationships between physical nature and society over time. It is argued that within each eco-cultural habitat, the relationship between physical nature and society is mediated by specific entanglements between technologies, institutions, and cultural values. These habitat-specific entanglements are neither ecologically nor culturally predetermined, but result from mutual adaptation based on variation (trial and error) and selection. It is shown how a variety of eco-cultural habitats evolves from this coevolutionary process. The book explores how these varieties come into being and how their specific characteristics affect the capacity to cope with environmental or social problems such as flooding or unemployment.

There are two case studies illustrating the potential of a coevolutionary understanding of the society–nature nexus. In the first, rural and urban settlement structures are conceptualized as distinct paths of eco-cultural adaptation. It is shown that each of these paths is characterized by predictable spatial correspondences between dwelling technologies, modes of social reproduction, cultural preferences, and related patterns in energy consumption (i.e. social metabolism). The second case study deals with flood protection in liberal and coordinated eco, welfare, and production regimes, drawing on lessons from the Netherlands and Hurricane Katrina in New Orleans. As a contribution to theory in environmental sociology, the coevolutionary perspective developed provides deeper insights into the intricate interplay between physical and social nature.

Johannes Schubert is a Research Assistant in the Department of Sociology at Ludwig-Maximilians University Munich, Germany.

Routledge Explorations in Environmental Studies

Environmental Adaptation and Eco-cultural Habitats

A coevolutionary approach to society and nature

Johannes Schubert

earthscan
from Routledge

Routledge
Taylor & Francis Group

LONDON AND NEW YORK

First published 2016
by Routledge
2 Park Square, Milton Park, Abingdon, Oxon OX14 4RN

and by Routledge
711 Third Avenue, New York, NY 10017

First issued in paperback 2017

Routledge is an imprint of the Taylor & Francis Group, an informa business

British Library Cataloguing in Publication Data
A catalogue record for this book is available from the British Library

Library of Congress Cataloguing in Publication Data
Schubert, Johannes, author.
    Environmental adaptation and eco-cultural habitats : a coevolutionary approach to
    society and nature / Johannes Schubert.
        pages   cm. – (Routledge explorations in environmental studies)
    Includes bibliographical references and index.
    1. Human beings–Effect of environment on.   2. Nature–Effect of human
    beings on.   3. Human ecology.   4. Habitat (Ecology)
    5. Coevolution.   6. Social evolution.   I. Title.
    GF51.S44 2016
    304.2–dc23   2015019633

ISBN 13: 978-1-138-49702-3 (pbk)
ISBN 13: 978-1-138-94285-1 (hbk)

Typeset in Times New Roman
by Out of House Publishing

Contents

Figures

Tables

Foreword

In this outstanding contribution – which is based on his PhD thesis at the Ludwig-Maximilians University of Munich, Germany in 2014 – Johannes Schubert explores in what ways correspondences between physical nature, technologies, institutions, and cultural preferences can result from coevolutionary mechanisms between physical nature and society. Such correspondences might also be described as elective affinities between natural circumstances and societal conditions. Landscapes selected for appropriate human inhabitants and human communities cultivate physical environs according to their ideational and cultural beliefs. But wait – isn't that eco-determinism or sociobiology (or even covert racism)? Shouldn't we have overcome "that" and freed social sciences from their paternalism by natural sciences, by now?

When taking ecological crisis as an opportunity to develop an interdisciplinary understanding of the society–nature nexus, it is important to scrutinize the historically grown self-conceptions of the disciplines involved. With the Enlightenment, the schism between humanities and natural sciences began. Materialism and rationalism claimed "progress" for themselves and – in doing so – separated themselves from what they believed to be "reactionary" religious thinking. All the same, humanities freed themselves from the stranglehold of religion in the meantime. At the beginning of the twentieth century, and as portrayed by Max Weber in his essays on epistemology ("Gesammelte Aufsätze zur Wissenschaftslehre"), these positions clashed in the neo-Kantian scientific dispute: while "free will" was believed to be the central subject of research in the humanities, it was held that natural sciences deal with nomologically determined objects. This dispute is still going on and shapes the scientific debate about different theoretical conceptions and methodological approaches: whereas proponents of an ideographic paradigm focus on singular descriptions, apologists of a nomological notion of science strive for regularities that can be generalized.

As observed by Max Weber, however, the alleged ontology of a "free mind" and a "determined nature" is misleading: although – in their self-descriptions – human individuals may claim to be guided by free will, from an external perspective it is often possible to predict the overall sum of human practices pretty accurately by statistical means. On this basis, social sciences emerged between the poles of humanities and natural sciences in the twentieth century. They scrutinize human behavior. However, they describe and explain human behavior not only from within, i.e. from a hermeneutical perspective, but also from the outside, i.e. with objectifying methods. Conversely, physical nature is not nearly as determined as physics classes in school would have us believe. The natural sciences of the eighteenth and nineteenth centuries draw their reputation – but also many misconceptions – from the fact that they idealized singular phenomena that can be easily predicted, such as the trajectories of the planets, to be "nature in general". Based on "scientific progress", there was also great hope that even such capricious processes as the daily weather would eventually comply with long-term forecasts. In the meantime, and beyond the backdrop of physics classes, natural sciences have further developed: they already go far beyond the idea of a purely determined object world, first with the theory of evolution, then with the theory of relativity, quantum physics, information theory, and – ever since powerful computers have been available – especially with the theory of complex systems (also referred to as "chaos theory").

Insofar, the dreams of never-ending technological progress seem to have come to an end, too: technological breakthroughs aiming at more advanced levels of controlling nature are becoming more and more expensive and – consequently – increasingly sparse. Moreover, they imply more and more unintended side effects. In this context, a scientific–political dilemma unfolds: if natural sciences took their own and most advanced theories seriously, they would have to qualify their promise of being able to exert control over nature as envisaged in the Enlightenment. In doing so, however, they would cut themselves off from abundant research funds. Thus, many natural scientists rather cling to an understanding of nature which strives for redemption in this world and is further compatible with the Enlightenment, physics classes, and the predominant technology policy. Irritating theoretical insights from their own discipline – which would claim more humility – are blocked out.

Consequently, and on the basis of advanced insights, there is no longer a theoretical foundation justifying an ontological schism between humanities and natural sciences. Physical nature affects human society and human society affects physical nature – in view of the ecological crisis we can no longer afford to turn a blind eye to this trivial insight of our common sense. In the same way as natural sciences let go their fantasies of omnipotence (at least theoretically), humanities no longer have to cherish defensive reflexes. Based on this awareness, a new notion of science has developed which underlies the Actor Network Theory (ANT) as represented, for example, by Bruno Latour, Annemarie Mol, John Law, and Isabelle Stengers. Here, physical nature and society are of equal status and both poles are granted initiative. Human and non-human actants form networks to change the world or to inhibit changes initiated by other networks. By following the idea that human and non-human entities group together in collectives, the ANT conceptualizes history as a dynamic in which both mind and matter "act" and in which what seemed to be irreconcilable differences between humanities and natural sciences no longer exist.

Principally, one can follow the ANT up to this point. When realizing, however, that the world is characterized not only by unpredictable but also by predictable phenomena, our world view should not only provide detailed descriptions of singular and contingent phenomena, but also abstraction and generalization. In the same way as it should be possible for ideographic natural history to exist along with nomological natural science, there should be room for social science – understood as a nomological discipline analyzing the interplay of human and non-human entities – alongside ideographic approaches to social history. By trying to establish a fundamentally new approach, the work of Johannes Schubert is located at this meta-theoretical level: physical nature and society are seen in connection, but not only on the basis of historical reconstruction (i.e. by ex post analysis as practiced by the ANT), but also – and this is where the main focus of this new perspective lies – on the basis of predictions. With this, the question of whether stable and ubiquitous regularities within configurations of human and non-human entities can be observed occupies center stage.

And indeed, some more or less surprising patterns seem to appear. Collectivists successfully run irrigated farming but fail when it comes to ordinary agriculture based on rain. And while cities yield a certain type of collective tolerance and open-mindedness, rural populations cultivate their conservative pertinacity. Here, the coevolution of mentalities and physical nature appears to be based on mutual selection: specific environs demand for certain adaptational technologies. These technologies, again, require institutional structures regulating human behavior which – in turn – necessitate suitable cultural preferences and mentalities to comply with and actively support these institutional structures and related behavioral expectations. If the book accounted only for this direction of influence, it could be blamed for eco-determinism, against which humanities – for good reason – have defensive reflexes. At this point of the discussion, however, the argumentative direction is changed: based on our cultural mentalities, we are able to look for and to create ideationally suitable institutions and to select and develop corresponding technologies in order to set out for new physical environs or to shape our physical surroundings according to our cultural needs and wants. Taken in isolation, this would be cultural determinism and blinkered as well. The innovative character of Schubert's work is its attempt to account for both perspectives in a symmetric and equal way. It is argued that these perspectives neither represent contradictions nor cancel each other out, but do ally with each other. Physical environs entail specific mentalities which – in turn – select and cultivate ideationally appropriate physical environs. In this process of mutual selection and adaptation, stable eco-cultural habitats emerge.

In Schubert's ideal-typical concept, there is only a limited number of eco-cultural habitats: there is one habitat based on individualism, low labor division, decentralized technologies, and dispersed settlement structures and one habitat in which collectivism, high degrees of labor division, centralized technologies, and compact settlement patterns correspond with each other. When considering the historic transition from agricultural to industrialized and urbanized economies and society, this is a convincing conception. Given the theoretical complexity of the present approach, the lowest possible number of types also favors a clear discussion. Nevertheless, and despite the fact that one should approach theory-building in an economical way, as a working group we are aware that a two-dimensional typology might be more expandable since it could account for the fact that industrial development can take place both under more individualistic and market-oriented as well as under more collectivistic and state-oriented institutional conditions. Likewise, both solitary and collaborative forms of work can be found in less developed societies.

Against the background of Chapters 2 and 3 in which the theoretical framework as sketched out above is elaborated, Schubert has conducted two case studies: settlement structures in Bavaria and South Tyrol (Chapter 5) and flood protection in river deltas (Chapter 6). With regard to settlement structures, Schubert's considerations are based on the biological theory of allometry: organisms reproduce themselves by extending or reducing their surface area, depending on whether they want to increase or decrease their metabolic exchange relations with their physical environment. Agriculture relies on collecting solar radiation and – for that reason – has to cover large areas of land. Industrial and (sub)urban economies, in turn, are based on labor division and internal exchange relations which are facilitated by short communication lines and hauls. Accordingly, distinct cultures of rural and urban ways of life have emerged over the course of time and can be observed today, even though an improved provision of infrastructure in rural areas and an increasing dispersion and suburbanization of cities can be observed. By using appropriate datasets, Schubert is able to show that strong spatial correspondences between physical environs, settlement patterns, social structures, and cultural preferences and mentalities exist. Moreover, and as described by fractal geometry, he is able to show that the observable differences between center and periphery exist on different spatial scales and – comparable to snow crystals – are nested in each other. With this consideration, his explanation – illustrated on the regional (Bavaria) and urban (Munich, Bolzano) scale – can be further expanded to the levels of private households or nation states.

Schubert's deliberations on flood protection are rooted in the perspective of comparative welfare state analysis. However, Schubert also covers aspects of the protection against environmental hazards. He can show that flood protection, like almost all other approaches to water regulation, requires cooperation and that institutional regimes characterized by welfare state solidarity are better prepared to cope with such issues. In contrast, liberal regimes rather tend to fail here, because egocentrism and legitimized inequalities only allow for individual and – as a consequence – often inefficient protection. Particularly for the poor, there is nothing left but escape, damage, or doom. Empirically, these considerations are illustrated by comparing historical approaches to flood protection in the Rhine and the Mississippi deltas (i.e. in the Netherlands and United States), whereby New Orleans and the Mississippi delta have not only been flooded by Hurricane Katrina, but had been hit several times before.

All in all, Schubert's work seems to represent a promising approach to do environmental studies in a symmetrical way, encompassing natural as well as social facts, understanding more contextual as well as more general patterns and developments. This is why I believe that this pioneering book merits a wide international audience.

Bernhard Gill
Munich, February 2015

Acknowledgments

A scientific work – at least in most cases – is not the product of an individual and isolated mind. It rather results from an intricate, long-lasting, and labor-intensive process embedded in the author's scientific and private environment. This is also true of the book presented here and I would like to express my gratitude to those persons and institutions that have essentially contributed to the progress of my work.

My special thanks go to my mentor Professor Bernhard Gill for his great support, his sensitive guidance, for all the inspiring discussions, the initiatives taken by him, and the encouragement that have come from his side.

I also give thanks to the German Federal Ministry of Education and Research (BMBF). Throughout the writing phase, I have been actively involved in ongoing research projects, first in the project "Klima regional" (BMBF-funding-initiative "Social dimensions of climate change and climate protection") and next in the project "Lokale Passung" (BMBF-funding-initiative "Environmentally friendly and socially acceptable transformation of the energy system"). The issues discussed in Chapter 5 as well as the data used there originate from these research projects. These and all related activities – teaching, conference attendances, holding presentations, or publishing papers – provided a crucial forum for sharing and critically discussing my ideas with a professional audience.

Finally, I would also like to thank my colleagues, Anna Wolff and Michael Schneider, as well as my friends and family for their support and encouragement.

Johannes Schubert
Kaufbeuren, February 2015

1 Introduction

The broader subject of this book is the relationship between physical nature and society. It attempts to describe and explain how human beings interact with one another and their physical environs (even if man-made). What can be observed is that human beings and human societies, respectively, are found in a variety of different, sometimes strongly contrasting physical environs. Unlike most animal or plant communities, human societies – thanks to social learning and cultural inheritance – are able to create their own artificial habitats in which they can survive. They have overcome practically all climatic and ecological conditions and spread out all over the world (and temporarily even beyond). In other words, human beings shape their physical surroundings in such a way as to serve their needs and wants. It is exactly here that the inquiry of this book starts. It firstly elaborates a theoretical frame of reference to help describe and understand the intricate interplay of physical nature and human society from a coevolutionary informed environmental sociological perspective. Secondly, this theoretical framework is exposed to first empirical tests.

There are a good number of examples of how varied the relationships between physical nature and society can be and of how they are mediated and take shape. They may become manifest in sedentarism and corresponding land use regimes but may just as well be reflected in patterns of seasonal migration. Observable adaptation strategies oscillate between low- and high-tech as well as between labor- and capital-intensive approaches; they range from solitary action to large-scale cooperation as well as from laissez faire-like conditions to highly centralized and institutionalized forms of coordination. Just take river deltas as an example: although they share crucial similarities such as low height above sea level (or even below) or land subsidence from a global perspective, it can be observed that distinct adaptation strategies are used to cope with volatile living conditions, ranging from floating gardens in the Ganges delta to massive dam-building along the Rhine–Meuse delta. However, these deltas are characterized not only by different technological solutions, but also by distinct institutional and cultural peculiarities. While floating gardens represent a low-tech and labor-intensive approach which can be individually and spontaneously made use of within family or clan, dam building and hydraulic engineering are high-tech and capital-intensive collectivistic adaptational strategies which heavily depend on hierarchic coordination and cooperation across space and time.

It is exactly this interplay between physical environs, different technological solutions, and corresponding institutional structures and cultural preferences which lies at the very heart of this book. It is argued that technological solutions alone do not suffice for environmental adaptation but that they have to be intertwined with fitting institutional and cultural structures to develop their full adaptational potential. Staying with the example of river deltas, how else could it be explained that New Orleans was repeatedly flooded whereas the Netherlands have mostly been spared – at least in the last 50 years and despite considerable similarities regarding physical-geographic conditions as well as technological expertise and solutions? Could it be that the predominant combination of collectivistic technologies such as dikes and liberally biased institutional settings and related mentalities in the Mississippi delta do not gear into each other as smoothly as in the Rhine–Meuse delta and its collectivistic approach to flood protection? Here, the central claim of this book is that environmental adaptation most crucially depends on suitable matches between physical nature, technologies, institutions, and cultural preferences. And while such matches coevolved in the Netherlands in the course of time, the collectivistic technology of dikes is implemented into an otherwise liberally biased institutional and cultural environment in New Orleans and the Mississippi delta. From the perspective of an external observer, suboptimal adaptation endeavors are the result, for instance the spatially unequal distribution of flood protection and related social problems.

To define the approach of the book still more precisely: it centers upon the fabric between physical nature and man as being rooted both in physical nature and culture. It searches for general principles, rules, and laws along which the ties of the fabric emerge. It analyzes how the resulting eco-cultural habitats develop over the years and how they turn into more and more rigid paths of eco-cultural adaptation in the course of time. It inquires the impact of physical environments on the selection of adequate adaptation technologies and the coming into existence of corresponding behavioral strategies and how all this translates into individualistic or cooperative cultural preferences. But also reversely, the present work addresses the effects of cultural preferences (or cultural path-dependencies) – once established – on the adaptation and coping capacities of eco-cultural habitats and their human inhabitants. It also discusses the question of how to account for the fact that – given comparable conditions – habitat-specific differences in approaching social and ecological issues can be observed.

To achieve these objectives, a theoretical frame of reference problematizing the relationship between physical nature and societies from a coevolutionary perspective is developed. When applying this frame of reference, what seems to emerge is a number of recurring patterns characterizing the relation between physical nature and society. It almost seems that specific physical environments, specific

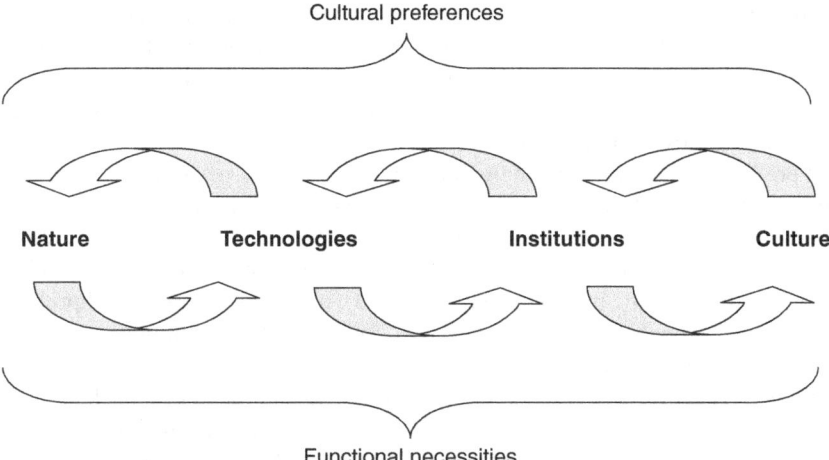

Figure 1.1 The eco-cultural fabric
Source: compiled by the author.

technologies and institutions, as well as particular cultural preferences amalgamate into habitat-specific elective affinities.

Thus, it is assumed that physical nature and society merge into particular matches constituting habitat-specific 'eco-cultural fabrics' (cf. Figure 1.1). The term eco-cultural fabric hints at the well-established term social fabric, which describes the structure of relationships between social actors. In the context of the present work, however, it appears to be more appropriate to operate with the term eco-cultural fabric: it is not restricted to social relationships, but relates these social textures to the properties of their surrounding physical environments. In doing so, it is argued that the relationship between physical nature and society is mediated by technologies, institutions, and hegemonic cultural value orientations. Thus, eco-cultural habitats (or eco-cultural paths of adaptation) are characterized by specific eco-cultural fabrics which – in turn – are defined as fit constellations between physical nature, technologies, institutions, and cultural preferences.

But how do these specific matches between physical nature, technologies, institutions, and cultural preferences actually come into existence? It is argued that eco-cultural habitats emerge in a coevolutionary way in the course of time. In other words, observable matches between physical nature, technologies, institutions, and cultural preferences are neither ecologically nor culturally predetermined, but result from mutual adaptation based on trial and error, selection, and variation. Society and nature – to a certain degree – simultaneously emerge and gain shape. This is also why well-adapted habitats and related eco-cultural fabrics usually account both for functional as well as ideational necessities. Otherwise, mismatches – that is, suboptimal adaptations – occur.

In order to gain a first impression of what a coevolutionary perspective actually is about, the concept of gene-culture coevolution and the example of lactose (in)tolerance are well suited for illustration. Basically, gene-culture coevolution addresses the dynamic interplay between genetic and cultural evolution (Henrich/ Henrich 2007, Bowles/Gintis 2011). "According to gene-culture coevolution, human preferences and beliefs are the product of a dynamic whereby genes affect cultural evolution and culture affects genetic evolution, the two being tightly intertwined in the evolution of our species" (Bowles/Gintis 2011: 14). Thereby, culture is defined "as the ensemble of preferences and beliefs that are acquired by means other than genetic transmission" (ibid.: 13), in which context social preferences "include a concern, positive or negative, for the well-being of others, as well as a desire to uphold ethical norms" (ibid.: 3).

A frequently cited example of gene-culture coevolution dates back to the Neolithic Revolution during which new cultural practices and farming techniques like sedentarism and dairy farming coevolved along with the ability of the human organism to digest lactose, especially in northern Europe (Henrich/Henrich 2007, Kallis/Norgaard 2010). Here, socially induced changes – that is, the change from a nomadic life form based on hunting and gathering to sedentarism – resulted in new selective and adaptational pressure regarding the physical nature (including human organisms). The unfamiliar conditions of sedentarism and dairying – particularly with regard to an increased dairy consumption – selected for human bodies able to digest lactose. In other regions, however, gene-culture coevolution led to a different outcome. Here, one and the same difficulty, namely digestive problems regarding dairy products, was solved by cultural techniques like processing milk into cheese. In other words, (slow-paced) biological evolution was outsmarted by cultural evolution.

> [I]n places such as the Middle East and China, the cultural transmission of the practice of keeping large domesticated animals was followed relatively rapidly by the evolution of the technological know-how for (and practice of) turning milk into cheese and yogurt. In these forms, lactase is not required, and anyone can obtain the nutritional benefits of milk. In these populations, adaptive cultural evolution beat natural selection acting on genes to the punch [...]
>
> Henrich/Henrich 2007: 31

Thus, changes in the social or physical nature do not remain where they occur, but exert selective pressure on any part of the eco-cultural fabric. Whether adaptation takes place in the social, physical (including human physiology), or in both environments is an empirical question.

Analogously, human societies can figuratively be conceptualized as organisms that adapt to physical nature by constructing man-made habitats in different physical (even if man-made) environments, for example, by building cities, houses, or roads and by adopting various cultural techniques like cooking or different degrees of cooperation and division of labor.

Socio-ecological coevolution involves a social niche construction; there is nothing fundamentally different between beavers and humans constructing water dams which in turn affect the evolution of various social subsystems such as water technologies, water institutions or consumption habits [...]

Kallis/Norgaard 2010: 692

Instead of genes, human societies are equipped with different technologies and organizational forms, various institutional structures, as well as cultural value orientations which are used to constitute and reshape physical nature according to human needs and wants (Norgaard 1994). And instead of selecting for genetic traits as in the case of lactose tolerance and individual digestion, physical nature, in turn, selects for fit matches between behavioral strategies, related technologies, appropriate institutional structures – in short culture – in the case of human societies.

While this perspective on the relationship between society and physical nature is part of the established body of knowledge in other disciplines – especially in ecological economics and anthropology, sociologists, with rare exceptions, rather shy away from explaining societal phenomena with reference to physical nature. In the sociological theoretical universe, Patrick Nolan and Gehard Lenski and their ecological-evolutionary theory make one of these exceptions. Their first premise is that "human societies are a part of the global ecosystem and cannot be adequately understood unless this fact is taken fully into account" (Nolan/Lenski 2011: 5). On this basis, Nolan and Lenski make the crucial point that the formation of the societal mode of organization – the formation of societies – represents an adaptive mechanism in its own right: "the societal mode of life became common in the animal kingdom for the same reason that wings, lungs, and protective coloring became common: They are all valuable adaptive mechanisms" (Nolan/Lenski 2011: 6).[1] Accordingly, human societies – just as any other species – are affected by changes in physical nature: "Because societies are adaptive mechanisms that mediate relations between a population and its environment, patterns of continuity and change in the environment necessarily affect patterns of continuity and change within societies" (Lenski 2005: 60).

Against the background of these considerations, physical nature and society are not treated as two opposed entities here, but are conceptualized as two mutually dependent constituents of an indivisible unit. Whenever a distinction is drawn between physical nature and society, this is an analytical differentiation. Thus, and in contrast to the dichotomous approaches which mostly predominate in environmental sociology, the coevolutionary perspective emphasizes that eco-cultural habitats are neither ecologically nor culturally predetermined, but result from mutual adaptation based on trial and error, selection, and variation (Norgaard 1997).

As argued by Lenski (2005: 83 ff.), the actual shape of eco-cultural habitats derives from the requirement of human beings to obtain energy and resources from their material surroundings. In his taxonomy of societies, observable variation in eco-cultural habitats is explained by the level of technological advance of a given society and the properties of its environmental surroundings: "each of the basic societal types reflects a unique combination of environmental and techno-logical characteristics and these determine what might be called their *operative technology*" (ibid.: 84). This procedure results in an idealized categorization of societies, such as hunting and gathering, herding, horticultural, maritime, or industrial societies.

While the author appreciates Lenski's overall commitment in establishing an evolutionary perspective in sociology, he does not agree with his somewhat narrow taxonomy of societies and the way in which his ecological-evolutionary theory is spelled out in detail, especially his techno-centric explanations and his fixation on subsistence technologies appear to be short-sighted and to require further elaboration. Also, the relationship between increasing industrialization, resource scarcity, and ecological crisis is not sufficiently addressed. Here, and from a global perspective, the looming end of fossil fuels and the fact that the unintended effects of industrialization cannot be relocated and externalized any further should be accounted for.

The present work has a somewhat different emphasis in explaining the emer-gence and operative logic of eco-cultural habitats. It concentrates on the inter-play between the existence of cooperative advantages (entailed with physical natures) and the emergence of operative cultures. To be more precise and follow-ing Thorstein Veblen (1990), the present work claims that there are two comple-mentary structural moments of habitat emergence: the ecological determination of culture and – once strong cultural preferences have been established – the cultural determination of physical nature and society.

According to these two structural moments, the emergence of eco-cultural habi-tats is explained in the following way: first, and with regard to the eco-deterministic explanation of culture (and respective technologies and institutions), it is argued that strong cultural preferences for cooperative action emerge when cooperative behavior is rewarded by physical nature (for example, in terms of more pleasure or higher calorie intakes) and when the benefits of cooperation outweigh its costs (for example, in the case of irrigation works or hydraulic engineering). Strong cultural preferences for solitary action, on the contrary, develop when the costs of cooperation outweigh its advantages, as, for example, in most forms of agri-culture (Ellickson 1993). Thereby, arguments from institutional economics can be harnessed to weigh the conditions under which the universally valid principles of labor division, economies of scale, or allometry make cooperation a benefi-cial behavioral strategy. In short, and depending on the occurrence of cooperative advantages, the first argument claims that physical nature selects for appropriate behavioral strategies which – in turn – entail fit technologies and related institu-tional structures and cultural preferences (cf. Chapter 2).

Second, it is elaborated that cultural preferences – once established and for rea-sons of higher cognitive and overall social efficiency – develop a life of their own and permeate more and more societal realms (cf. Chapter 3). These cultural guid-ing heuristics represent general rules of conduct regulating humans' interference with each other as well as humans' interference with physical nature within one and the same adaptational path. Now, cultural preferences – in turn – select for socio-ecological challenges (and related technologies and institutions) which can be processed within the respective path-specific cultural bias. "Sometimes elements of culture are preserved, not because they are superior solutions to problems, but simply because they ensure standardized behavioral responses in situations where these are essential" (Nolan/Lenski 2011: 46). Thus, the central argument is that predominant cultural biases select for ideationally appropriate matches between themselves and physical nature, technologies, and institutions. In the course of time, this process results in cultural specialization and related cultural path-dependencies (Douglas/Wildavsky 1982, Bednar/Page 2007).

With regard to the adaptive capacities of eco-cultural habitats, cultural specialization goes hand in hand with path-specific advantages and vulnerabilities. Thus, the argument ends with some critical reflections on the Janus-faced character of cultural specialization which, for example, becomes apparent when cultural self-stabilization takes place at the expense of functional necessities and mismatches occur. "When the beliefs and values involved are felt to be sufficiently important, a society may reject the most economic solution to its needs in favor of a solution that is ideologically preferable" (Nolan/Lenski 2011: 59).

Another crucial aspect of cultural specialization is that eco-cultural habitats are characterized by internal self-similarities as described by, for example, Benoit Mandelbrot's fractionalization theory (Mandelbrot 1967, 1987). Snowflakes are a famous example. Studying them under the microscope, it becomes apparent that a single snowflake is composed of many ice crystals which totally look alike and feature equal proportions. In other words, self-similar objects show the same geometric or static properties on different scales. Transferring these observations to eco-cultural adaptation, it is claimed that similar habitat-specific characteristics – especially with regard to the matches between physical nature, technologies, institutions, and cultural preferences and related coordinative principles – can be found on different analytical scales, be it nation states, municipalities, cities, rural villages, or households.

What results from this approach can be described as a theoretical continuum on which all paths of eco-cultural adaptation can be located. From an ideal-typical perspective, this continuum expands between two poles: eco-cultural paths of adaptation characterized by strong preferences for solitary action, individual freedom, and autonomy (and corresponding technologies and institutional structures) on the one side and eco-cultural paths of adaptation based on cooperative behavior coevolving together with collectivistic cultural preferences and related technologies and institutional structures on the other.

For the purpose of this book – which is, above all, dedicated to theory building – the following theoretical discussion of eco-cultural habitats as well as respective illustrations and case studies concentrate on consciously chosen paradigmatic examples. Thus, the somewhat pointed character of the empirical illustrations and case studies is intended – first, in order to illustrate central theoretical arguments and second, to demonstrate their applicability. For sure, this procedure does not do justice to the empirical variance in eco-cultural paths of adaptation – but this is not the objective of this book. Rather, it aims to introduce a coevolutionary perspective into the theoretical repertoire of environmental sociology.

Quite legitimately, the reader could ask why the approaches of environmental sociology and ecological economics should not suffice to conceptualize the relationship between human society and nature. After all, both disciplines claim for themselves to be specialized in the topic of mankind–environment relationships. So, why not leave everything the way it is? What can be gained from bringing together theoretical arguments from environmental sociological and ecological economics under the umbrella of a coevolutionary perspective? Isn't that just a dubious analogy and an inappropriate theoretical transfer from biological evolution to social processes as, for example, indicated by Winder et al. (2005)?

Without calling the theoretical and methodological achievements of environmental sociology and ecological economics into question, both disciplines appear to be limited by specific blind spots. While the rigid front between the apologists of a materialistic perspective on the one side and a culturalistic perspective on the other has led to a deadlock obstructing a sufficient understanding of the environment–society nexus in environmental sociology, ecological economics has made good progress in theorizing and modeling the material basis of social life – albeit with a somewhat limited notion of society and social action. In the author's view, these are the unfortunate results of the former fragmentation of social sciences. A critical theory of society, however, has to overcome both reductionisms – the limited sociological understanding regarding the material basis of social life as well as the limited economic understanding of social processes. How far these objectives can be achieved by a coevolutionary perspective is one of the central objectives of this book. At best, this could result in an ecological economically informed environmental sociology as well in a sociologically informed ecological economics. Without doubt, this is an ambitious project and the present work understands itself only as a first step.

These charges and claims as well as the suggested remedy (i.e. a coevolutionary perspective) are not new. Most prominently, Richard Norgaard argued for "A coevolutionary environmental sociology" in *The International Handbook of Environmental Sociology* in 1997 (also see Norgaard/Kallis 2011). One year later, the editors of the very same handbook – Graham Woodgate and Michael Redclift – stood up for a transformation "From a 'sociology of nature' to Environmental Sociology: Beyond Social Constructivism" in 1998:

> We suggest a more balanced view of the relationship between society and its underlying material and natural conditions. We must move beyond the position where nature is viewed as *either* the material conditions of our existence, *or* as no more than a set of culturally generated symbols. We must begin to accept nature as both.
>
> Ibid.: 7

However, it appears that their voices were more or less unheard – at least in the environmental sociological community (Manuel-Navarrete/Buzinde 2010).

All the same, Lenski's (2005) approach of an ecological-evolutionary theory of society has not been taken note of, yet. Thus, environmental sociology is still caught in the unproductive dualism of an either constructivist or materialist perspective.

Apologists of the social-constructivist camp emphasize the important fact that humans' interference with physical nature is not predetermined by ecological conditions, but crucially depends on social construction and the attribution of meaning and sense (Berger/Luckmann 1966, Douglas/Wildavsky 1982, Schwarz/Thomson 1990). The material basis of social life is not taken into account in most cases. In contrast, apologists of the materialist camp emphasize the important services ecosystems fulfill for human beings as well as for society as a whole. They call attention to the fact that social life – beyond all forms of social constructivism – ineluctably depends on a physical-material basis which essentially contributes to its formation (Catton/Dunlap 1978, Dunlap/Catton 1979, Grundmann/Stehr 1997). Both perspectives make strong assumptions about the relationship between physical nature and society, whereby one of the two is generally taken for granted and is thought to shape the other in a rather unidirectional or even deterministic way (Norgaard 1994: 75 ff., Woodgate/Redclift 1998, Dunlap 2010). For that reason, and despite their particular epistemological merits, both perspectives represent insufficient analytical tools when the dynamic interplay between physical nature and society occupies center stage.

Addressing the relationship between physical nature and society from a coevolutionary perspective, however, it is claimed that these shortcomings can be dealt with. "At an epistemological level coevolution offers a powerful logic for transcending environmental and social determinisms and developing a cross-disciplinary approach in the study of socio-ecological system" (Kallis/Norgaard 2010: 690). Within a coevolutionary informed theoretical framework, the hitherto conflicting positions of the materialist and culturalist perspective can be related to one another in a meaningful and productive way – the eco-determination of culture and the cultural determination of nature are complementary forces in the emergence of eco-cultural habitats. Because different readings and concepts of coevolution exist, the specific notion of coevolution used here should be further clarified.

In the broadest sense, coevolutionary approaches usually deal with the explanation of change between two or more entities over space and time. Giorgos Kallis and Norgaard (2010) distinguish between the following coevolutionary mechanisms and related fields of research: biological, social, gene-culture, bio-social, and socio-ecological coevolution. Trying to locate the approach of eco-cultural coevolution as developed here within these fields of research, it lies at the interface of social and socio-ecological coevolution. Social coevolution deals with the reciprocal evolution of social systems. To give an example, it is frequently used to develop a deeper understanding of how and why specific technologies coevolve together with specific behavioral strategies and suitable institutional structures (van den Bergh/Stagl 2003).

A typical example is the case of flood protection and social welfare in the Netherlands (cf. Chapter 6.2.1). Socio-ecological coevolution in turn – "refers to cases where evolution in the social system affects the bio-physical environment, which, in turn, affects evolution in the social system" (Kallis/Norgaard 2010: 692). Typical examples are Norgaard's work about the coevolution of socio- and ecosystems in the Amazon (Norgaard 1981) or Donald Worster's elaborations on the emergence of the Dust Bowl and its socio-ecological consequences (Worster 1979, cf. subchapter 3.2)

With this, the concept of eco-cultural coevolution developed here basically follows the notion of coevolution as represented by Norgaard (Norgaard 1994, Kallis 2007).

> In the coevolutionary paradigm, the environment determines the fitness of people's behavior as guided by alternative ways of knowing, forms of social organization and types of technologies. Yet, at the same time, how people know, organize and use tools determines the fitness of characteristics of an evolving environment. At any point in time, each determines the other. Over time, neither is more important than the other. And depending on genetic mutations, value shifts, technological changes and social innovations that arise randomly, the evolutionary path is reset for a period until another change occurs.
>
> Norgaard 1997: 163

Kallis summarizes Norgaard's notion of coevolution as follows: "For Norgaard coevolution is a process of coupled change between practices, values and the biophysical environment. Humans change environments both materially and cognitively, he argued, and in turn new environments change human practices and ideas" (Kallis 2007: 1). Two systems "coevolve when they have a causal impact on each other's evolution" (Kallis 2007: 1 ff.). Evolution, in turn, is defined as "a process of selective retention and renewable variation [...] Evolution involves the three Darwinian processes of variation, inheritance, and selection" (Kallis/Norgaard 2010: 690).

Although most authors agree that the coevolutionary logic offers an epistemological entry-point "to transcend dichotomous debates that confound EE [ecological economics, author's note] and environmental studies" (Kallis/Norgaard 2010: 692), the concept of coevolution represents a controversially discussed issue nevertheless (Hodgson 2010). With regard to the approach of eco-cultural coevolution as proposed here, the following two issues appear to be relevant: first, the extent to which the logical structure of evolution is the same between social and biological phenomena is controversially discussed (Winder 2005, Winder et al. 2005, Hodgson 2004, Kallis 2007, Hodgson 2010). In particular, Winder critically questions the transferability of a Darwinian theoretical framework to social and cultural evolution. Instead, he argues for an understanding of cultural evolution that follows from a Lamarckian point of view or the "Darwin-Huxley synthesis". This also raises ontological issues about the use of a coevolutionary vocabulary. Here again, Winder et al. (2005) call for a narrower and more careful use of biological concepts and terminology. Complex eco-dynamics do not necessarily fulfill the conditions for coevolution in a strict biological sense and – for that reason – should rather be described as co-dynamic or mutually dependent when the conditions of coevolution are not met. Second, coevolutionary approaches are often accused for biological reductionism, a mechanistic notion of society (Hodgson 2010) and "a deep structuralist bias that assumes a rigid causal determination in social life" (Manuel-Navarrete/Buzinde 2010: 139).

Regarding the first issue, the relevant literature appears to support a notion of coevolution as represented by Norgaard (Hodgson 2004, 2010, Lenski 2005, Kallis 2007, Nolan/Lenski 2011). Above all, this is the case because there are more similarities than differences between social and biological evolution.

> In recent decades, our understanding of biological evolution and sociocultural evolution has advanced dramatically. It is now clear for the first time that *both types of evolution are based on records of experience that are preserved and transmitted from generation to generation in the form of coded systems of information.* In the biological evolution, the record of experience is preserved and transmitted by means of the genetic code. In sociocultural evolution, the record is preserved and transmitted by means of symbol systems. Both the genetic "alphabet" and symbol systems provide populations with the means of acquiring storing, transmitting, and using enormous amounts of information on which their welfare and, ultimately, their survival depend. Thus, *symbol systems are functional equivalents of the genetic alphabet.* It is hardly surprising, therefore, that there is a fundamental similarity in the way the two evolutionary processes operate. Both processes involve random variation and selective retention.
>
> Nolan/Lenski 2011: 61

Having these fundamental similarities in mind, a theoretical transfer from Darwinian biological to cultural evolution appears to be a legitimate procedure which is also advocated by Hodgson (2004, 2010). Further, the present work shares the basic tenor of these authors that much more can be gained from a holistic, integrative, and interdisciplinary comprehensible meta-theoretic Darwinian framework à la Norgaard than from a stricter and narrower theoretical and terminological approach as urged for by Winder et al. (2005).

Following Lenski's (2005: 6 ff.) distinction between older and newer evolutionary theories, the second issue – the charge of biological reductionism and a mechanistic notion of society – can be invalidated. In general, the older approaches from biology and social sciences struggled with insufficient causal explanations: while the biological approaches "had no adequate explanation for the process of variation" (ibid.: 7), social sciences "lacked a satisfactory explanation of the cause of societal development and growth" (ibid.: 7). Among other things, this provided a breeding ground for racism and biologism. In contrast, the newer approaches from biology and social sciences are much more advanced.

Lenski (2005) summarizes the advantages and theoretic achievements of modern evolutionary theory as follows: modern evolutionary theory emphasizes:

> (1) humanity's common genetic heritage, (2) the various technologies our species has fashioned to enhance this heritage, (3) the resources and constrains of the biophysical environment, (4) the resources and constrains of the socio-cultural environment, and (5) the impact of the process of inter-societal selection.
>
> Ibid.: 7

Thus, eclectic or racialist explanations of the older evolutionism are replaced by more complex and better informed theoretical approaches. Further, and in contrast to the old evolutionary theories, modern evolutionism rejects its normative faith in the idea of progress (Norgaard 1994) and its deterministic explanations of social and cultural evolution. Finally, "the newer evolutionism rests on a firmer and richer foundation of archaeological, ethnographic, and historical data" (Lenski 2005: 8, also cf. Wilson 2010). In view of these developments, there is no necessity to cherish general suspicions against evolutionary and coevolutionary approaches.

From a sociological perspective, caution is called for nevertheless: as indicated above, coevolutionary approaches from ecological economics tendentially operate with a somewhat mechanistic or reductionist notion of society. To mitigate against these potential limitations, David Manuel-Navarrete and Christine Buzinde (2010) suggest that environmental sociologists "should bring to the fore of the debate the universe of meanings, creativity and bewilderment that characterize cognitive systems and human agency" (ibid.: 139). By claiming that the formation of eco-cultural habitats is neither ecologically nor culturally determined, this is exactly what characterizes the approach of eco-cultural adaptation: it not only accounts for the ecological determination of culture, but also strengthens a social-constructivist perspective on the cultural determination of physical nature. It is the very dynamic of these two complementary and sometimes conflicting forces which actually characterizes eco-cultural coevolution and the emergence of distinct adaptational paths.

Against the background of these critical considerations – what can be gained from the concept of eco-cultural coevolution? First, and similar to Bruno Latour (1987, 1993), the concept of eco-cultural coevolution promises to solve the issues that go hand in hand with unidirectional approaches as sketched out above. Consequently, the coevolutionary perspective promises to provide deeper insights into the intricate relationship between physical nature and society.

Second, and in contrast to Latour, who overcomes the imbalances of ecological and cultural determinism by means of detailed historical and context-specific descriptive analyses of micro-sociological phenomena, the approach of eco-cultural coevolution has the potential of a universally valid and generalizable theory yielding strong explanatory power and predictive capacity on different analytical levels. These claims derive from two facts: first, it is argued here that the coming into existence of eco-cultural habitats is decisively characterized by the existence of cooperative advantages, especially in terms of labor division, economies of scale, and allometric rules, which – in turn – can be accounted for by what might be called historically invariant quasi-natural laws. Thus, unlike other coevolutionary approaches, the concept developed here is not restricted to explanations referring to evolutionary and historically contingent events and path-dependencies, but also suggests that historically invariant mechanisms exist which can be used to explain eco-cultural adaptation. Second, and with regard to habitat-specific self-similarities, the explanatory power and predictive capacity of the present approach are not limited to the micro level, but can be applied to different scales, be it nation states, cities, communalities, or households.

Third, the approach of eco-cultural coevolution presents a theoretical linkage facilitating the connection of theoretical currents from various scientific fields, especially from environmental sociology (Douglas/Wildavsky 1982, Kraemer 2008), environmental psychology (Schwartz 1970, 1994, Stern et al. 1995, Stern 2000), as well as from evolutionary (van den Bergh/Stagl 2003, Bowles 2004, Kallis/Norgaard 2010), environmental (Olson 1965, Ostrom 1990), institutional (North 1990), behavioral (Henrich et al. 2001), and cultural economics (Bednar/ Page 2007). Thus, the present work understands itself as an open and integrative theoretical frame of reference which aims to conceptualize the relationship between physical nature and society beyond the well-trodden paths of environmental sociology.

The book proceeds as follows: the theoretical considerations regarding the coevolutionary emergence of eco-cultural paths of adaptation as sketched out above will be further elaborated in Chapters 2 and 3 and summarized in Chapter 4.

In the two subsequent chapters, this theoretical frame of reference will undergo its first empirical tests. Therefore, the following questions particularly occupy center stage: is it possible to operationalize the concept of eco-cultural adaptation on different analytical scales and to apply it to both quantitative and qualitative research questions? Does it yield the theoretically expected results regarding habitat-specific correspondences between physical nature, technologies, institutions, and cultural value orientations? Do the selected examples actually show the theoretically expected self-similarities? Moreover, the case studies presented in Chapters 5 and 6 lend themselves toward illustrating the synthesis of environmental sociology and ecological economics as indicated above.

By intertwining the approach of eco-cultural adaptation with the concept of social metabolism as described by Rolf Peter Sieferle et al. (2006), rural and urban modes of life are conceptualized as specific paths of eco-cultural adaptation in Chapter 5. Whereas rural paths of adaptation initially emerged under the influence of the agrarian metabolic regime based on the collection of solar energy, urban paths of adaptation, industrialization, and related modes of life became an extensive phenomenon only when energy demands could be met by fossil fuels. Accordingly, the eco-cultural fabrics of rural and urban paths of adaptation should show significant differences. Chapter 5 shows that these differences can be captured empirically by quantitative cross-sectional analyses and that some matches between technologies, institutions, and culturally biased ways of life – more frequently than by random chance – appear in rural than in urban settlement structures (and vice versa). Thereby, and due to habitat-specific self-similarities, the leading assumption is that typical and recurring patterns between physical nature, technologies, different modes of social reproduction, and related cultural preferences can be found on different spatial scales. In other words, a snapshot identifying contemporary constellations between specific physical environments, technologies, institutions, and cultural preferences in different settlement structures will be carved out from quantitative empirical data.

Based on these considerations, working hypotheses regarding the spatial correspondences between physical nature (for example, rural or urban areas), dwelling technologies (for example, building types or heating systems), cultural preferences (for example, in terms of political orientation, familialism, traditionalism), and related modes of social reproduction (measurable by, for example, different degrees of labor division, childlessness, or female employment rates) are formulated and tested on different analytical scales. For that purpose, Chapter 5 is structured in the following way: by making use of (individual) data from the German Socio-Economic Panel, spatial patterns in dwelling technologies and related patterns of energy consumption (in terms of carbon dioxide emissions) are identified for Germany in a first step in subchapter 5.1. Based on these nationwide analyses, census-like (aggregated) data from all 2056 Bavarian municipalities are used in subchapter 5.2 to analyze whether the spatial patterns in dwelling technologies and carbon dioxide emissions identified in subchapter 5.1 actually correspond with distinct cultural preferences and modes of social reproduction in the theoretically expected ways. Last but not least, and using self-collected household data from Munich (Bavaria) and Bolzano (Italy), subchapter 5.3 aims to show that the technological and cultural differences characterizing rural and urban modes of life and related metabolic properties can also be found within urban settlement structures.

Chapter 6 deals with the questions of how eco-cultural paths of adaptation take shape in the course of time and – once established – affect the coping capacities of their human inhabitants. These questions are approached from a qualitative and historic-reconstructive perspective which makes use of a paradigmatic comparison of two deltas and the respective eco-cultural paths of adaptation: the Rhine–Meuse–Scheldt and the Mississippi deltas. This research approach was motivated by the observation that social and ecological issues – especially in the realms of social welfare and flood protection – are dealt with differently, even though both deltas arguably share key similarities.[2] Despite these similarities, history shows that the Netherlands have mostly been spared severe flooding for more than 50 years now, whereas the devastation of New Orleans by Hurricane Katrina in 2005 revealed that the city – once again – was badly prepared to cope with extreme weather and its humanitarian aftermath. So, how can these differences be explained?

The central idea of Chapter 6 is that different approaches to social and ecological challenges can be explained by path-specific forms of institutional and cultural specialization which – in turn – should reflect in distinct (mis)matches between physical nature, technologies, institutions, and cultural preferences. Seen from this perspective, the devastation of New Orleans by Hurricane Katrina and its humanitarian aftermath is nothing exceptional but just offers another occasion to study the structural mismatches between a liberally biased mode of life on the one hand and the institutional and cultural necessities associated with the provision and maintenance of public goods such as flood protection on the other. Or to put it in more general terms: it is assumed that non-excludable goods such as flood protection suffer in liberally biased institutional and cultural settings but bloom in eco-cultural habitats characterized by strong preferences for cooperative behavior and hierarchic coordination.

Given this central research idea, Chapter 6 proceeds as follows: in a first step, and by using the relevant literature about different welfare and production regimes as well as corresponding forms of environmental regulation, different regulatory approaches coordinating the utilization of physical nature and societal processes are identified and paradigmatically compared. In doing so, it is shown that both deltas are overlaid by different regulatory environments organizing the utilization of physical nature and societal process in distinct ways (subchapter 6.1). Based on this institutional inventory, subchapter 6.2 is devoted to the following questions: what can be said about the emergence of these path-specific institutional peculiarities? How did they mold past and still mold present approaches to ecological and social challenges? And – last but not least – how do they affect individual resources to cope with extreme weather events such as Hurricane Katrina? Answering these questions, the emergence of path-specific institutional and cultural specialization is illustrated against the background of flood protection in the Netherlands. The ambivalent effects of institutional and cultural specialization on habitat-specific adaptation capacities are illustrated against the example of New Orleans and its environs.

The book concludes with some critical reflections gauging the work's aspirations against its findings and with a short outlook on possible future research topics. Despite some ambivalent empirical results, it is concluded that the concept of eco-cultural adaptation meets aspirations as sketched out above: it presents an analytical frame of reference able to describe and explain the intricate relationship between physical nature and society from both a materialist and culturalist perspective. It proves to be operationalizable – both in quantitative and qualitative terms and on different analytical scales – and yields the theoretically expected results, at least in most cases. Taken together, these findings suggest that the approach of eco-cultural adaptation presents itself as a universally valid theory rather than as a heuristic merely applicable to a limited number of cases. However, more research and modeling is clearly necessary to better understand the dynamic character of eco-cultural paths of adaptation – all the more so because it is only applied to the Western world and related eco-cultural fabrics. Additionally, the theoretical range of the approach of eco-cultural adaptation, its predictive capacity, and its explanatory power should be further assessed – all the more so as two different epistemes explaining the emergence of eco-cultural habitats and

their operative logics are used here: the evolutionary paradigm relying on historically contingent events and path-dependencies on the one side and the universally valid principles of labor division, economies of scale, and allometry on the
other. It is not clear yet, however, in what way these epistemes are related to each
other and the nature of their specific contributions to the coming into existence of
eco-cultural habitats and their operative logics.

Notes

1 Please note that Lenski follows biological taxonomy, according to which human societies are part of the animal kingdom (Lenski 2005: 33).
2 Both deltas are characterized by comparable geological peculiarities such as low elevation above sea level (or even below) and land subsidence, have comparable technological
   know-how and financial resources at their disposal, and employ similar technologies –
   levees, ditches, sluices, water pumps, and the like – to create and protect habitable space.

Bibliography

Bednar, J., & Page, S. (2007). Can Game(s) Theory explain culture? The emergence of
    cultural behavior within multiple games. *Rationality and Society*, 19(1), 65–97.
Berger, P., & Luckmann, T. (1966). *The social construction of reality; a treatise in the
    sociology of knowledge* (1st ed.). Garden City: Doubleday.
Bowles, S. (2004). *Microeconomics: Behavior, institutions, and evolution.* Princeton: Princeton
    Univ. Press.
Bowles, S., & Gintis, H. (2011). *A cooperative species*. Princeton: Princeton Univ. Press.
Catton, W. R., & Dunlap, R. E. (1978). Environmental sociology: A new paradigm. *The
    American Sociologist*, 13(1), 41–49.
Douglas, M., & Wildavsky, A. (1982). *Risk and culture: An essay on the selection of
    technical and environmental dangers*. Berkeley: Univ. of California Press.
Dunlap, R. E. (2010). The maturation and diversification of environmental sociology: from
    constructivism and realism to agnosticism and pragmatism. In M. R. Redclift & G.
    Woodgate (Eds.), *The international handbook of environmental sociology* (pp. 15–32).
    Cheltenham: Edward Elgar.
Dunlap, R. E., & Catton, W. R. (1979). Environmental sociology. *Annual Review of
    Sociology*, 5(1), 243–273.
Ellickson, R. (1993). Property in Land. *Faculty Scholarship Series,* Paper 411. http://
    digitalcommons.law.yale.edu/fss_papers/411
Grundmann, R., & Stehr, N. (1997). Klima und Gesellschaft, soziologische Klassiker und
    Außenseiter. Über Weber, Durkheim, Simmel und Sombart. *Soziale Welt*, 48(1), 85–100.
Henrich, J., Boyd, R., Bowles, S., Camerer, C., Fehr, E., Gintis, H., & McElreath, R. (2001).
    In search of Homo economicus: Behavioral experiments in 15 small-scale societies. *The
    American Economic Review*, 91(2), 73–78.
Henrich, N., & Henrich, J. (2007). *Why humans cooperate: A cultural and evolutionary
    explanation*. Oxford: Oxford Univ. Press.
Hodgson, G. M. (2004). Darwinism, causality and the social sciences. *Journal of Economic
    Methodology*, 11(2), 175–194.
    (2010). Darwinian coevolution of organizations and the environment. *Special
    Section: Coevolutionary Ecological Economics: Theory and Applications*, 69(4),
    700–706.

Kallis, G. (2007). When is it coevolution? *Ecological Economics*, 62(1), 1–6.

Kallis, G., & Norgaard, R. B. (2010). Coevolutionary ecological economics: Special Section: Coevolutionary ecological economics: Theory and applications. *Ecological Economics*, 69(4), 690–699.

Kraemer, K. (2008). *Die soziale Konstitution der Umwelt*. Wiesbaden: VS Verlag.

Latour, B. (1987). *Science in action*. Cambridge: Harvard Univ. Press.

(1993). *We have never been modern*. Cambridge: Harvard Univ. Press.

Lenski, G. E. (2005). *Ecological-evolutionary theory: Principles and applications*. Boulder: Paradigm.

Mandelbrot, B. (1967). How long is the coast of Britain? Statistical self-similarity and fractional dimension. *Science*, 156(3775), 636–638.

(1987). *Die fraktale Geometrie der Natur*. Basel: Birkhäuser.

Manuel-Navarrete, D., & Buzinde, C. N. (2010). Socio-ecological agency: from 'human exceptionalism' to coping with 'exceptional' global environmental change. In M. R. Redclift & G. Woodgate (Eds.), *The international handbook of environmental sociology* (pp. 136–149). Cheltenham: Edward Elgar.

Nolan, P., & Lenski, G. (2011). *Human societies: An introduction to macrosociology*. Boulder: Paradigm.

Norgaard, R. B. (1981). Sociosystem and ecosystem coevolution in the Amazon. *Journal of Environmental Economics and Management*, 8(3), 238–254.

(1994). *Development betrayed: The end of progress and a coevolutionary revisioning of the future* (1st ed.). London, New York: Routledge.

(1997). A coevolutionary environmental sociology. In M. Redclift & G. Woodgate (Eds.), *The International Handbook of Environmental Sociology* (pp. 158–168). Cheltenham: Edward Elgar.

Norgaard, R. B., & Kallis, G. (2011). Coevolutionary contradictions: prospect for a research programme on social and environmental change. *Geografiska Annaler: Series B, Human Geography*, 93(4), 289–300.

North, D. C. (1990). *Institutions, institutional change, and economic performance*. Cambridge: Cambridge Univ. Press.

Olson, M. (1965). *The logic of collective action; public goods and the theory of groups*. Cambridge: Harvard Univ. Press.

Ostrom, E. (1990). *Governing the commons; the evolution of institutions for collective action* (1st ed.). Cambridge: Cambridge Univ. Press.

Schwartz, S. H. (1970). Moral decision making and behavior. In L. Berkowitz & J. Macaulay (Eds.), *Altruism and helping behavior* (pp. 127–141). London: Academic Press.

(1994). Are there universal aspects in the structure and contents of human values? *Journal of Social Issues*, 50(4), 19–45.

Schwarz, M., & Thompson, M. (1990). *Divided we stand: Redefining politics, technology, and social choice*. Philadelphia: Univ. of Pennsylvania Press.

Sieferle, R. P. et al. (Ed.) (2006). *Das Ende der Fläche: Zum gesellschaftlichen Stoffwechsel der Industrialisierung*. Cologne: Böhlau.

Stern, P. C. (2000). Toward a coherent theory of environmentally significant behavior. *Journal of Social Issues*, 56(3), 407–424.

Stern, P. C., Dietz, T., Kalof, L., & Guagnano, G. A. (1995). Values, beliefs, and proenvironmental action: Attitude formation toward emergent attitude objects. *Journal of Applied Social Psychology*, 25(18), 1611–1636.

van den Bergh, J. J., & Stagl, S. (2003). Coevolution of economic behaviour and institutions: towards a theory of institutional change. *Journal of Evolutionary Economics*, 13(3), 289–317.

Veblen, T. (1990). *The place of science in modern civilization and other essays.* New Brunswick: Transaction Publishers [1919].

Wilson, D. S. (2010). *Darwin's cathedral: Evolution, religion, and the nature of society.* Chicago: Univ. of Chicago Press.

Winder, N. (2005). Modernism, evolution and vaporous visions of future unity: Clarification in response to Norgaard. *Ecological Economics*, 54(4), 366–369.

Winder, N., McIntosh, B. S., & Jeffrey, P. (2005). The origin, diagnostic attributes and practical application of co-evolutionary theory. *Ecological Economics*, 54(4), 347–361.

Woodgate, G., & Redclift, M. (1998). From a 'Sociology of Nature' to environmental sociology: Beyond social construction. *Environmental Values*, 7(1), 3–24.

Worster, D. (1979). *Dust Bowl.* Oxford: Oxford Univ. Press.

2 The ecological determination of behavioral strategies and cultural preferences

In focusing on functional necessities regarding the fits between physical nature, technologies, institutions, and culture, this chapter exclusively deals with the eco-deterministic explanation of cultural preferences. However, the emergence of cultural preferences does not directly result from the properties of physical nature, but happens in response to the formation of fit behavioral strategies which – for their part – result from the natural environment. Thereby, it is argued that all paths of eco-cultural adaptation can be located on a continuum which – from an ideal-typical perspective – expands between the behavioral strategies of solitary versus cooperative action. Whereas non-cooperative behavioral strategies entail the emergence of cultural preferences such as individual freedom and autonomy, cooperative behavioral strategies coevolve together with collectivistic cultural preferences. But under what conditions and according to what rules do behavioral strategies of solitary or collective action and related cultural preferences emerge and prevail?

As argued by Veblen (1990) and Brette (2003), "individuals start developing habits of action, under the impulse of their instincts, and in contact with the material conditions of the society they live in" (Brette 2003: 460). Depending on whether the properties of the physical environment reward solitary or cooperative action, respective behavioral strategies and related technologies, institutional structures, and cultural preferences will emerge in the course of time. Thereby, and as argued by Samuel Bowles and Herbert Gintis (2011), a "high ratio of benefits to costs makes cooperation an evolutionarily likely outcome because [...] the likelihood that a population will develop and maintain cooperative practices is higher, the greater are the benefits of cooperation" (ibid.: 197). In other words, when the benefits of cooperative behavior outweigh its costs,[1] cooperation – "engaging with others in a mutually beneficial activity" (ibid.: 2) – and respective cultural preferences will emerge.

Starting their analysis in the Late Pleistocene, Bowles and Gintis (2011) explain the emergence of cooperative behavior by the functional advantages that are derived from division of labor and positive returns to scale:

> Our Late Pleistocene ancestors inhabited the large-mammal-rich African savannah and other environments in which cooperation in acquiring and

sharing food yielded substantial benefits at relatively low costs. The slow human life-history with prolonged periods of dependency of the young also made the cooperation of non-kin in child rearing and provisioning beneficial. As a result, members of groups that sustained cooperative strategies for provisioning, childrearing, sanctioning non-cooperators, defending against hostile neighbors, and truthfully sharing information had significant advantages over members of non-cooperative groups.

Ibid.: 3

Envisioning prehistoric men hunting down big game or protecting each other against predators, it becomes apparent that cooperative behavior – as opposed to solitary action – turned out to present the superior behavioral strategy.

However, Bowles and Gintis (2011) also make the point that "cooperation is not an end, but rather a means. In some settings, competition, the antithesis of cooperation, is the more efficient means to a given end" (ibid.: 5). Whether solitary action or cooperation proves to be the better adapted behavioral strategy depends on the particular properties of a given eco-cultural fabric and remains an empirical question. Whatever the case may be, there is no doubt that the physical environment selects for suitable behavioral strategies and appropriate (cultural) technologies. Cooperative action goes hand in hand with the emergence of "connected" technologies (that is, technologies which depend on frequent and long-term cooperation), for example, irrigation systems. It is the other way round when solitary action represents the superior behavioral strategy. In this case, non-cooperative behavioral strategies and "unconnected technologies" – defined as technologies which can individually and spontaneously be applied by individuals or small groups, such as hunting small game and gathering – are likely to coevolve.[2]

In this context, the approach of eco-cultural coevolution as presented here claims that suitable behavioral strategies in terms of solitary action or cooperation do not exist on their own, but entail the emergence of functionally adequate cultural preferences. Cultural preferences are generally described as collectively shared basic values which emerge in the course of social interaction. Here, the presence of cooperative advantages crucially shapes the properties of social interaction which, in turn, affects the importance of reputation building and the emergence of related social norms and cultural preferences (Henrich et al. 2004). Commonly shared cultural preferences foster mutual understanding, enhance expectability and – for these reasons – reduce transaction costs on a daily basis. In doing so, cultural preferences institutionally stabilize and ideologically legitimize predominant behavioral strategies.

In the case of Lamalera whale hunters, for example, the practice of dividing the prey takes place in accordance with strict rules reflecting cultural values emphasizing the overall importance of sharing:

In real life, when a Lamalera whaling crew returns with a whale or other large catch, a specially designated person meticulously divides the prey into predesignated parts allocated to the harpooner, crew members, and others participating in the hunt, as well as the sailmaker, members of the hunters' corporate group and other community members (who make no direct contribution to the hunt).

Henrich et al. 2004: 39

Here, the functional necessity of cooperative hunting gave rise to cultural preferences and related daily routines (and rituals) which ensure that hunting and sharing whales will be played like a repeated cooperative game (from a game theoretical perspective). In other words the Lamalera whale hunters do not "share valued resources because they have social preferences, but they have social preferences because they live in a place where hunting whales is the best way to make a living, and those who hunt large game do better if they learn how to share" (Bowles/Gintis 2011: 13).[3]

Trying a short summary, one might say: not only do physical environments select for fit behavioral strategies, they also select for corresponding and functionally adequate cultural preferences. With this, the concept of eco-cultural coevolution also adopts central ideas from organizational sociology and institutional economics according to which specific labor processes, goods, or technological applications should be matched with appropriate (business) cultures (for example, in order to reduce transaction costs or to motivate workers to invest in (firm-)specific knowledge or other assets). Simplifying this somewhat, the following typology of tasks and technologies on the one hand and appropriate (business) cultures and governance structures on the other can be derived from the relevant theoretical literature (for example, Olson 1965, Ouchi 1980, Douglas/Wildavsky 1982, Williamson 2000, 2003, Hall/Soskice 2001).

As summarized in Table 2.1, it is assumed that strong cooperative advantages – in the long run – entail collectivistic cultural preferences (and related daily routines and practices) whereas cooperative disadvantages result in solitary action and rather entail individualistic value orientations. Supportive evidence can also be found in the work of Joseph Henrich et al. (2004). Conducting game theoretic experiments like the Ultimatum Game in different small scale communities, they were able to show that cultural preferences heavily depend on potential payoffs to cooperative behavior and the degree of market integration (ibid.: 28 ff.). In the case of the Machiguenga, a semi-nomadic social group primarily depending on horticulture, the payoffs to (non-kin) cooperation are very low – a constellation which is accompanied by strong individualistic value orientations. In contrast, the previously mentioned Lamalera whale hunters heavily depend on cooperative behavior and show strong (non-kin) altruistic cultural preferences.

From a strictly eco-deterministic and functional perspective, such coincidences should result from the fact that only well-adapted functional matches between physical nature, behavioral strategies, and cultural preferences have a chance to prevail. Functionally suboptimal matches (for example, strong individualistic value orientations combined with the functional necessity of cooperatively hunting whales) are sorted out via variation and selection in the course of time. As pointed out by Henrich et al. (2004), that should be the case because the presence

Table 2.1 Eco-determination of behavioral strategies and cultural preferences: the example of rainfed agriculture and irrigation

Environmental conditions	High precipitation, humidity	Low precipitation, aridity
Appropriate agricultural adaptational strategy	Rainfed agriculture	Irrigation works and irrigated farming
Type of adaptational action and related technologies and goods	Physically and socially unconnected tasks and technologies, e.g. plowing or seeding	Physically and socially connected tasks and technologies, e.g. construction and maintenance of irrigation systems
Cooperative advantages and type of behavioral strategy	Low Adaptation individually and spontaneously achievable, low paybacks to reputation building → Solitary action	High Adaptation requires collective effort and long-term perspective, high paybacks to reputation building. → Collective action
Type of transaction cost-economizing governance structure	Ranges from laissez-faire/ self-subsistence to market-coordination	Ranges from clan-like governance structures to hierarchic coordination.
Adequate property rights/rights of disposal	Individual rights of disposal	Collective rights of disposal
Compatible cultural preferences	Individual freedom and autonomy, solidarity limited to kin	Solidarity and tolerance, in modern societies even among strangers

Source: compiled by the author.

of cooperative advantages crucially affects the properties of social interaction, reputation building, and the emergence of related social norms and cultural preferences.

> The structure of social interaction affects the benefits and costs of reputation building and other relationship-specific investments and thereby alters the evolution of common norms and the degree of social ties. Societies differ markedly in the frequency of interaction with known individuals and the degree to which interactions are governed by complete contracts as opposed to informal guarantees related to trust and reputation. We know from experiments, for example, that trust and interpersonal commitment often arise where contracts are incomplete, but not under complete contracting […]. If trust and commitment are important parts of one's livelihood, these sentiments may be generalized to other areas of life or evoked in situations which appear similar to everyday life.
>
> (Ibid.: 46 ff.)

But why should all this be so? What theoretical arguments could justify the obser-vation that cooperative disadvantages and solitary action entail individualistic value orientation whereas cooperative advantages are reported to be accompanied by collective action and community spirit? The following considerations on the occurrence of cooperative advantages or disadvantages (in terms of high coord-ination costs) reveal first theoretical answers, which are further elaborated on the basis of the paradigmatic examples of rainfed agriculture and irrigated farming (and hydraulic engineering).

To begin with, Bowles and Gintis (2011) define cooperation as "engaging with others in a mutually beneficial activity" (ibid.: 2). From the perspective of eco-nomic functionalism, it can be argued that mutual benefits can result from division of labor, economies of scale, and allometric physical laws (which represent a sub-set of the economies of scale). With this, mutual benefits first and foremost result from gains in productivity, saving of costs, and higher resource efficiency.

However, cooperation with others does not necessarily turn out to be a mutu-ally beneficial activity, but – in many constellations – becomes a very costly endeavor. That is the case because defining and transferring rights to and obliga-tions of those who cooperate represents a key element of collaboration which – depending on the respective goods or services involved – can entail intensive bargaining procedures as well as costly arrangements to control and enforce informal agreements or formal contracts (North 1990). This may be the case when measuring the quality and quantity of goods or services in an intricate endeavor (for example, due to low measurability or asymmetries of information). In contrast to complete contracting (which means that all rights and obligations regarding a given property object as well as its quality, quantity, and other char-acteristics can be explicitly defined by all parties to the exchange), incomplete contracting demands alternative institutional governance structures to ensure that all parties involved will fulfill their obligations (Demsetz 1967, North 1990, Williamson 2000, 2003). Therefore, and contrary to anonymous (spot-market) exchange relations where buyers and sellers only have to meet once in order to accomplish the exchange of, for example, toothbrushes or apples via complete contracting, incomplete contracts result in exchange relations where exchange partners have to meet frequently for the sake of renegotiation and reputation building in order to compensate for low measurability by trusted relationships.[4]

In this context, Douglas North (1990) describes institutions as "humanly devised constraints that shape human interaction" (ibid.: 3) in order to cope with such insecurities:

> The costliness of information is the key to the costs of transacting, which consist of the costs of measuring the valuable attributes of what is being exchanged and the costs of protecting rights and policing and enforcing agreements. These measurement and enforcement costs are the source of social, political, and economic institutions.
>
> North 1990: 27

Depending on the overall complexity of exchange relations, North (1990) distinguishes three different types of exchange relations. As a general rule of thumb, he implies that the higher the degree of specialization and labor division (i.e. the more people get involved), the more intricate the coordination of exchange relations becomes. Thus, and depending on the respective level of specialization and related degrees of complexity, North (1990) identifies the following institutional structures which – in the course of time – emerge in order to stabilize cooperation: in the case of low labor division which primarily involves "small-scale production and local trade", highly personalized exchange relations based on repeated dealing, "cultural homogeneity (that is a common set of values), and a lack of third-party enforcement (and indeed little need for it) have been typical conditions" (ibid.: 34). With rising degrees of specialization and labor division and related necessities such as long-distance trade between different cultures, sheer clientelism does not suffice any longer. To define and enforce all relevant rights and obligations amongst the parties involved in such situations, "a second general pattern of exchange has evolved, that is impersonal exchange, in which the parties are constrained by kinship ties, bonding, exchanging hostages, or merchant codes of conduct" (ibid.: 34 ff.). Last but not least, impersonal exchange and cooperation amongst strangers (expanding time and space) can also be facilitated by third-party enforcement (for example, by the Leviathan).

The present section started out with the assumption that cooperative advantages can result from universally valid rules as described by the principles of labor division, economies of scale, and allometric physical laws. In this context, North's considerations above highlight the key fact that cooperative advantages (and their utilization) are not self-evident but heavily depend on either relationships of trust or third-party enforcement (for example, by the Leviathan). Thus, third-party enforcement and relationships of trust represent similar means to one and the same end, namely to cope with information problems and – in doing so – to enable the utilization of cooperative advantages. Against the background of these fundamental considerations, the properties of labor division, economies of scales, and allometry will be discussed next. In this context, the question of how cooperative advantages come into being is of special interest.

As illustrated by Bowles' and Gintis' (2011) example of prehistoric men, members of cooperative groups had significant advantages over members of non-cooperative groups. These advantages resulted from division of labor, economies of scale, and allometry.

Cooperation in acquiring and sharing food or in child rearing and provisioning are classical examples of labor division. Here, cooperative advantages arise while optimizing the ratio between working time and output via task-specific specialization (Smith 1776). Accordingly, workers:

> do better by specializing in subsets of the tasks, and then combining their outputs with that of other workers who specialize in other tasks. The increasing

returns from concentrating on a narrower set of tasks raises the productivity
of a specialist above that of a jack-of-all-trades.

Becker/Murphy 1992: 1139

However, there are clear limits to the advantages of specialization (Becker/
Murphy 1992). First, division of labor and specialization only make sense when
specialized skills (and respective goods) can be exchanged (for example, within
families, firms, or in the market place). Robinson Crusoe, isolated hermits, or
nomads clearly do better as "jack-of-all-trades" than, for example, highly specialized surgeons or sociologists. Second, and even if markets to exchange specialized skills exist, more and more specialization does not automatically result in
higher degrees of productivity. As emphasized by Gary Becker and Kevin Murphy
(1992), specialization and division of labor are limited by the costs and intricacies
of coordinating specialized workers. "Modern work on principal-agent conflicts,
free-riding, and the difficulties of communication implies that the cost of coordinating a group of complementary specialized workers grows as the number of
specialists increases" (ibid.: 1138). As long as gains in productivity due to specialization outweigh coordination costs, division of labor and cooperation present
the superior behavioral strategy.

Defense against hostile neighbors or predators is another example of how
cooperative behavior paid off for prehistoric men. Here, cooperative advantages
are not only derived from division of labor as described above, but also from
economies of scale. Joaquim Silvestre (1987) defines economies of scales like
this: "We consider the unit costs of producing a (single or composite) output
under a given technology (no technical change). We say that there are *economies*
(or *diseconomies*) *of scale* in some interval of output if the average cost is decreasing (or increasing) there." With this, economies of scale can be described as cost
advantages that occur due to increases in quantity in the broadest sense (Bowles
et al. 2005: 384). The general argument goes like this: given high fixed costs (for
example, investments in protective barriers like palisades or levees), average costs
per unit decrease with increasing numbers of people and livestock that are shel-
tered (Mosca 2006). The same holds true, for example, for infrastructure facilities
such as pipes for fresh- and wastewater or gas (network technologies): average
costs per meter decrease with increasing numbers of households that are connected (cf. Figure 2.1).

However, there also are limits to economies of scale. Regarding the example
of protective barriers, average costs per unit do not infinitely fall with increasing numbers, but may rise when exceeding the limits of capacity (for example,
when too many people seek shelter within protective barriers and interfere with
each other in emergencies or when too many households flush their toilets at
the same time). In addition, those who cooperate have to establish some rules
of the game (North 1990) defining, monitoring, and enforcing rights and obligations regarding the utilization and maintenance of protection barriers or other
infrastructure facilities. As long as the costs of coordinating such common

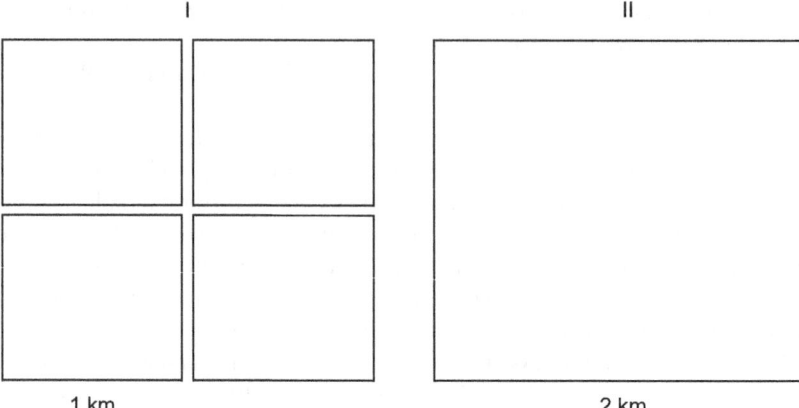

Imagine four square kilometers of diked land (a square with two kilometers on each side). In the first case, each square kilometer of land is separately embanked, meaning that each single dike has an entire length of four kilometers. Thus, four square kilometers of embanked land add up to a total dike-length of 16 kilometers. In the second case, a single dike embanking the whole square is built. The total length only adds up to eight kilometers, indicating huge saving potentials regarding the amount and costs of labor, time, and materials necessary to construct and maintain the embankment.

Figure 2.1 Economies of scale
Source: compiled by the author.

facilities do not exceed their benefits, cooperation is the superior behavioral strategy. More generally, this poses the question of optimal size, for example, of firms or (non-)governmental bureaucratic organizations. In this context, Oliver Williamson (1967) argues that diminishing returns to scale are particularly due to losses of control which occur when firms or organizations become so large that meaningful coordination across the hierarchy levels turns out to be virtually impossible.

> For any given span of control [...] an irreducible minimum degree of control loss results from the simple serial reproduction distortion that occurs in communicating across successive hierarchical levels. If, in addition, goals differ between hierarchical levels, the loss of control can be more extensive.
>
> Ibid.: 135

Last but not least, cooperative advantages can result from allometric physical laws as described by the square-cube law (Galilei 1730), dealing with the relationship of surface area and volume of three-dimensional objects such as prehistoric longhouses: squaring the surface of a given object results in its volume increasing cubically (cf. Figure 2.2). In the case of building physics and energy consumption, for example, this means that large buildings have comparatively smaller surfaces and – for that reason – waste less energy than single family houses.[5] Biological research about the mass-to-metabolism ratio also shows that large animals need

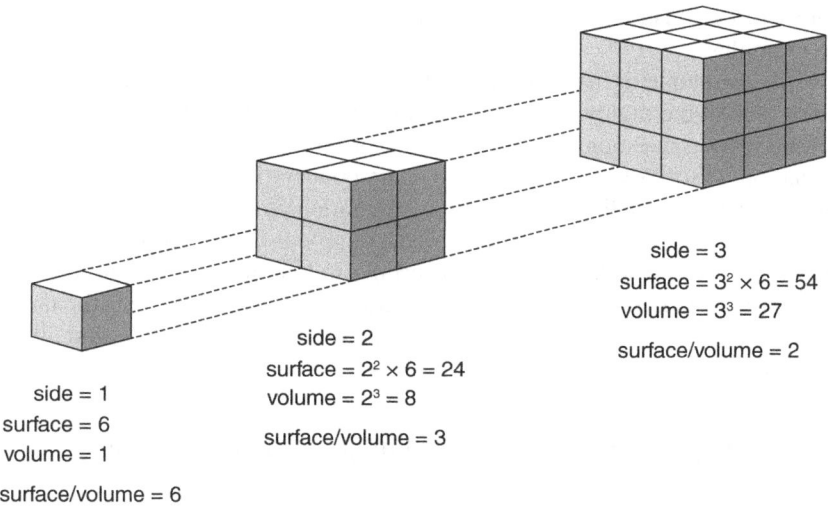

side = 3
surface = $3^2 \times 6 = 54$
volume = $3^3 = 27$

surface/volume = 2

side = 2
surface = $2^2 \times 6 = 24$
volume = $2^3 = 8$

surface/volume = 3

side = 1
surface = 6
volume = 1

surface/volume = 6

Figure 2.2 Surface-to-volume ratio
Source: compiled by the author.

comparatively less energy per unit of volume than smaller ones (Schmidt-Nielsen 1984, Vogel 1988, West et al. 1997).

Luis Bettencourt et al. (2007) tried to prove similar regularities for cities and the urban metabolism. In accordance with the square-cube law and the mass-to-metabolism rate, the authors show that similar economies of scale can also be shown for cities, but only with regard to their material infrastructure.[6] In contrast to mammals, however, it is reported that the urban metabolism does not slow down but accelerates with increasing city size.[7]

Against the background of these considerations, cooperative advantages can be derived from physical and social concentration. In the case of physical concentration (for example, in the form of longhouses instead of solitary and widespread huts), cooperative advantages become manifest in higher resource efficiency (for example, with regard to building materials or heating energy). In the case of social concentration (for example, via household size), cooperative advantages appear as saving of costs as described by economies of scale. However, and analogous to the limits of labor division and the limits of economies of scale, physical and social concentration result in cooperative advantages only within given capacity limits and as long as coordination efforts do not become rampant.

By now, it should have become apparent that the ratio of cooperative advantages to related costs of coordinating cooperation determines which behavioral strategy will prevail. Prevailing behavioral strategies, in turn, go hand in hand with appropriate technologies and cultural preferences. In the course of time, matches between physical environments, behavioral strategies, technologies, and cultural preferences become more and more intertwined and gradually condensed into institutional structures and ideologies. How these processes actually operate will be illustrated in the following chapters by means of two paradigmatic examples.

The first example is derived from the realms of rainfed agriculture and revolves around the highly disputed question of the optimal operational size of farms and related questions about the degree of internal labor division. Here, a combination of solitary action (or small-scale cooperation mainly on the level of family farms) and individualistic value orientations seems to characterize the most appropriate adaptive path. The second example is derived from the realms of irrigated farming and hydraulic engineering. It illustrates why a combination of cooperative behavior and collectivistic cultural preferences presents the most appropriate adaptive strategy to operate dams, levees, drainage canals, and flood protection.

2.1   Rainfed agriculture and the advantages of solitary action

In his seminal paper "Property in Land", Robert Ellickson (1993) pursues the objective of deriving "positive and normative propositions about the evolution of land regimes" (ibid.: 1320). In footnote 14, he specifies this intention:

> Because land-tenure arrangements are basic, many analysts treat them as exogenous independent variables when attempting to explain other phenomena, such as economic productivity, social equality, and the like. In this Article, a land regime is treated as a dependent variable that is affected by technologies, scale efficiencies, risk, ideologies, and other variables regarded as independent.
>
> Ibid.: 1320

With regard to Table 2.1, land regimes can thus be described as adaptational strategies which deterministically result from physical nature and the occurrence of cooperative disadvantages which – in turn – entail functionally appropriate technologies and institutional and cultural superstructures. With this, the specific properties of a given land regime heavily depend on the question of whether functional advantages can be derived from cooperation, for example, in the form of higher productivity due to specialization and division of labor, economies of scale, allometry, or the sharing of risks (for further reading see Abel 1967, Dahlman 1980, Ellickson 1991, and Schulze 2007).

In accordance with the above considerations on cooperative advantages and the nature of transaction costs, Ellickson (1991, 1993) argues in support of an efficiency thesis which "asserts that *land rules within a close-knit group evolve so as to minimize its members' costs*" (Ellickson 1993: 1320), defining such a group as "a social entity within which power is broadly dispersed and members have continuing face-to-face interactions with one another" (ibid.: 1320).[8] According to Ellickson (but also to North 1990, Ostrom 1990, Williamson 2000, 2003, and Ostrom/Walker 2003), such group conditions are conducive to cooperation because they provide each member "with both the information and opportunities they need to engage in informal social control" (ibid.: 1320 ff.) in the absence of third-party enforcement. With this, land regimes as described by Karl Dahlman (1980) or Ellickson (1991, 1993) represent transaction cost-economizing governance structures which – from a functional perspective – emerged in such a way as to minimize behavioral strategies such as shirking or free-riding and related costs of monitoring and enforcement.

To be more precise, Ellickson (1993) discerns between two different transaction cost-economizing land regimes: regimes that developed to deal with "small events" on the one hand and regimes that came up in order to deal with "large events" on the other. Whereas small events such as growing tomatoes do not affect others and can be achieved individually, large events such as fighting bush-fires or providing and maintaining irrigation systems greatly affect all members of a community and can only be achieved in combined efforts. Therefore, small events can be described as being socially and physically unconnected whereas large events can be characterized as being socially and physically connected.

In the case of small events such as growing tomatoes, the costs of cooperation are higher than its benefits, meaning that individual property in land should emerge as the transaction cost-economizing governance structure in the course of time. Above all, the high costs of coordination (and related cooperative disadvantages) can be explained by the fact that group ownership in land is highly prone to free-riding, shirking, and grabbing. Applied to the tomato example, group ownership in land means that all community members have equal rights and obligations regarding the cultivation of the tomato plants and the harvest. That – in turn – results in a situation in which each community member would be better off if he or she could capture his or her share of the harvest without contributing to its cultivation. Furthermore, group ownership also means that all community members are allowed access to the common tomato garden – a fact which turns effective boundary control and monitoring into an intricate and confusing endeavor if grabbing is to be avoided. In short, governing tomato gardens via group property would result in substantial welfare losses.

> A group may be able to devise internal institutions for coping with these problems, but [...] these mechanisms are likely to be much more costly than the simple monitoring systems associated with individual land ownership. From a transaction-cost perspective, a commune faces a choice between the Scylla of endless evening meetings and the Charybdis of an ever-increasing pile of unwashed dishes in the sink.
>
> Ellickson 1993: 1348 ff.

Against the background of these considerations, Ellickson (1993) argues in favor of property regimes based on individual ownership in land, suggesting that they simplify monitoring tasks and provide incentive structures which ensure the responsibilities for cultivation and border control rest on those who have strong personal incentives to conscientiously fulfill these tasks because they fully bear the profits (or losses) of personal commitment (or laziness). To begin with, Ellickson points out that "self-control [...] is much simpler than the multiperson coordination entailed in intragroup monitoring" (ibid.: 1327). Second, he argues that detecting trespassers and potential grabbers "is much less demanding than evaluating the conduct of a person who is privileged to be where he is" (ibid.: 1327). Finally, individual ownership goes hand in hand with strong incentives motivating the individual owner to seriously carry out boundary protection or any other monitoring tasks. Actually, a "sole owner bears the entirety of any loss stemming from slack oversight, whereas a group member bears only a fraction" (ibid.: 1328). According to Ellickson (1993), these arguments also explain why "family farming is ubiquitous, why collectivized agriculture almost always fails, and why virtually no dwelling units are shared by groups as large as 25" (ibid.: 1331 ff.).

Recalling the previous considerations on possible explanations for the coincidence of cooperative disadvantages and the emergence of functionally adequate value orientations, it does not come as a surprise that strong cultural preferences for individual freedom and autonomy accompany private property regimes in land, for example, in North America (Flinn/Johnson 1974, Dalecki/Coughenour 1992, Beus/Dunlap 1994). The presence of cooperative advantages or disadvantages crucially molds the properties of social interaction, the importance of trust and reputation building, and – in doing so – the emergence of appropriate norms and cultural preferences which institutionally stabilize and ideologically legitimize the properties of social interaction. Because, in most cases, the costs of coordination do not outweigh the benefits in the case of farming, individualistic value orientations constitute the predominant and coordination cost-economizing set of cultural preferences supporting individual property in land and solitary action.[9]

However, individual ownership in land potentially generates new tradeoffs. Above all, this is the case because individual property goes hand in hand with parcelling out land in comparably small pieces which – to be sure – reduces coordination costs but impedes efficiencies of territorial scale (cf. Figure 2.1) and generates new transaction costs. First, and with regard to boundary control and economies of scale, "the costs of fencing and other forms of perimeter monitoring drop per acre enclosed. This mathematical relationship has prompted many traditional societies to graze live stock on expansive group-owned pastures" (Ellickson 1993: 1332). Second, and as the persistence of plantation growing shows, a "large territory also permits a landowner to use more specialized equipment and workers and to marshal gangs of workers for projects for which returns to scale exist" (ibid.: 1332). In the case of individual property in land, however, individual owners cannot profit from such effects. Third, rights of way and procedures dealing with, for example, cattle trespassing have to be negotiated, monitored, and enforced (Ellickson 1991). Taken together, however, the overall costs apparently do not justify group ownership yet. Instead, these costs are tolerated in order to avoid the significantly higher coordination costs associated with group ownership in the case of small events.

Last but not least, an interesting exception should be briefly discussed: (temporary) advantages of group ownership in land regarding positive returns to territorial scale and the sharing of risks. Using historical evidence from pioneer settlements in North America – namely Jamestown, Plymouth, and Salt Lake – Ellickson (1993) reports the interesting phenomenon that in "each case, the settlers initially opted for group ownership, but after a few years switched to private ownership of intensively used lands" (ibid.: 1336). For explanation, it is helpful to consider what settling on unknown soil actually means. In the beginning, settling is primarily accompanied by high degrees of uncertainty and precariousness and demands for high initial investments in common infrastructure and goods such as palisades, streets, or irrigation systems (large events). In addition, there is neither state nor market to provide insurance against a diverse set of risks such as illness or bad harvest. Although trade companies initially financed settlers and partly stepped in during emergencies, those companies often were not within reach, meaning that settlers primarily were left on their own. Given such conditions, group ownership in land represented a functionally appropriate solution to exploit returns to territorial scale and to benefit from labor sharing (for example, in the case of large events such as defending against tribal raids or constructing palisades and irrigation works).

However, the early settlers did not differentiate between large and small events, meaning that farmland was also held as a collective asset. Whereas group ownership turned out to be beneficial in the case of palisades or irrigation (due to economies of scale and allometric principles), group ownership in land – for the reasons discussed above – resulted in severe famine and even death (Ellickson 1993: 1343). For Jamestown, and despite the fact that settlers were starving, it is even reported that they shirked from cultivating food or going hunting (activities which would contribute to the common good of the whole community) but preferred to bowl in the streets, instead (ibid.: 1337). Due to this significant mismatch, settlers successively began to convert governance structures regulating the cultivation of food. This process eventually resulted in farmland regimes based on private ownership.

> In sum, after a few years, the risk spreading benefits of group land ownership would no longer outweigh its familiar shortcomings [...]. At that point, the settlers understandably would switch to private land tenure, the system that most cheaply induced individuals to accomplish small and medium events that are socially useful.
>
> Ellickson 1993: 1342 ff.

In the case of large events such as irrigation works, however, group ownership and centralized regulation prevailed, as in the case of the Mormon settlers at Salt Lake. Therefore, the following chapter deals with the question of why (and how) group ownership – despite its well-known intricacies – prevailed in the case of providing and maintaining irrigation works.

2.2 Irrigation farming and the advantages of cooperation

The fascinating thing about Salt Lake Valley is that initially only Mormon settlers were able to permanently colonize these unfavorable regions by cooperatively building and maintaining irrigation works whereas other settlers usually failed (Arrington/May 1975, Arrington et al. 1976). Against the background of the present perspective according to which cultural preferences are determined by physical nature, a plausible explanation would be that the physical properties of Salt Lake Valley – namely low annual precipitation and severe drought – selected for specific technologies, suitable behavioral strategies, and related cultural preferences. Seen from this angle, the physical properties of Salt Lake Valley selected for behavioral strategies and cultural preferences uniquely featured by Mormon pioneers. The Mormons' success in cooperatively building and maintaining irrigation works could be interpreted as an unintended byproduct stemming from internal group characteristics and governance structures which had never been intentionally designed to serve the purpose of coordinating irrigation works.

Referring to the Mormon community as "The Lord's Beavers" and terminologically inspired by Karl August Wittfogel (1957), Worster (1985) provides first key empirical and theoretical explanations for the Mormons' success:

> Though they came to the arid West with empty pockets and a lack of training, the Mormons did have a system of hierarchy and group discipline, and that critical quality made possible their rapid success in water manipulation. [...] In the ancient irrigation states a pattern of hierarchy had often evolved slowly, century by century, out of the intensification of irrigation. In Utah the organization was more exogenous, but it proved to work as well as any Mesopotamian order of priests and kings in achieving river control. It provided a unified scheme of development and – within the limits of the available technology – a maximization of resource exploitation, and it freed the communities from individuals squabbling over water rights. It allowed the amassing of capital to undertake new projects and provide a cushion of security when projects failed. And, most important, it claimed to speak with the voice of God.
>
> Worster 1985: 77 ff.

Whereas behavioral strategies of ordinary pioneers were primarily guided by the Jeffersonian ideal of yeoman farmers as embodied by the Homestead Acts fostering individual property in land and solitary action, Mormon settlers were rooted in strong collectivism and hierarchic governance structures urging them towards cooperative behavior. In the Republicans' world view of the late eighteenth and early nineteenth centuries, yeoman farmers – independent family farmers making a living from subsistence or commercial agriculture – were seen as the central constituent of society. Under the impression of feudalistic European serfdom, property in land was considered as a key precondition for individual freedom and political autonomy. This fundamental attitude is also reflected in the Homestead Acts which granted parcels of land (usually 160 acres) to individual homesteaders at little or even no cost (Elkins/MacKitrick 1993, Foner 1995). While non-Mormon pioneers – on average – were equipped with nothing more than their individual wisdom, the labor force of a few family members (or the members of their trek), and some scarce resources such as tools or money to cope with the arid West (strong constraints regarding individual capacity and budget), the Mormon Church organized the foundation of new colonies in tactical ways and provided their settlers with multisided moral, ideological, strategic, and material support (Arrington et al. 1976, Worster 1985). Thus, Mormon communities were embedded in a cultural and institutional environment characterized by collectivistic value orientations and favorable organizational properties such as centralized planning, internal labor division, and the sharing of risks (whereas ordinary settlers in most cases had nothing). Last but not least, the Mormon Church

> asserted that all water ultimately belonged to the commonality, not to the autonomous individuals. That notion, a radical one by the standards of mid-nineteenth-century America, was justified by the argument that in such an arid place, where water was scarce and survival easily put at risk, a single authority must have ultimate jurisdiction over its disposal. And what more trustworthy authority was there than the Church?
>
> Worster 1985: 78

Before going into further theoretical detail here, it pays to have a short look at what happened when the West, including the Mormons' territory, was annexed by the federal government and the Mormon Church gradually lost its influence on irrigation.

In the late nineteenth century, "the state of Utah changed its laws to allow, for the first time, private individual ownership of the water resource. A key element in Mormon communalism was thus destroyed" (Worster 1985: 82). Wherever private property rights in water were adopted, severe problems arose: "Encouraging as it did a fevered, competitive race to exploit, the doctrine led straight to a chaotic war of claims against counterclaims" (ibid.: 92), and settlers began to relocate their farms further and further upstream in order to appropriate and extract as much water as possible. Consequently, farms located further downstream were no longer able to properly irrigate their fields (Coman 1911, Worster 1985). Obviously, private property in water (interpreted as a mechanism to coordinate the allocation of scarce water resources) resulted in significant mismatches, giving rise to governmental irrigation acts, determined to "find a collective counterforce to chaotic individualism. They were the first steps toward declaring that the rivers of the West are in some sense public property and that any private appropriation can only be made under public rules and at public sufferance" (Worster 1985: 95). With this, centralized governance structures (as initially established by the Mormon Church) regulating the provision and allocation of water became re-established. This can be interpreted as another empirical hint that only sufficiently adapted fits between physical properties, technology, institutions, and cultural preferences have a chance to prevail in practice. Moreover, the case of "chaotic individualism" and respective governmental remedies underlines the theoretical considerations according to which eco-cultural paths are always characterized by the complementary forces of functional necessities and cultural preferences.

Now, and against the background of these empirical illustrations and considerations, a more theoretically focused discussion should elaborate why cooperation is advantageous in the case of irrigation works and scarce water resources and how Mormon pioneers were able to cope with problems of collective action such as free-riding and shirking on a day-to-day basis.

From the perspective of economic functionalism, cooperative behavior (as opposed to solitary action) presents the superior strategy to provide and maintain irrigation works. As a result of the universally valid laws of labor division, economies of scale, and allometry, this is the case because irrigation works (as well as any other physically and socially connected network technology) are characterized by high payoffs to cooperation – at least as long as the costs associated with coordinating cooperation do not become rampant. The costs per unit associated with building and maintaining ditches or dikes decrease with increasing numbers of fields connected to one and the same irrigation canal. Accordingly, the costs each settler has to bear, for example, in terms of labor force, time, and investments in machinery, decrease with increasing numbers of participants (which also solves the problem of high start-up investments). Because water resources were scarce and irrigation works – sooner or later – were subject to capacity constraints, water was conceptualized as a common good by the Mormon Church in order to prevent overexploitation or unequal rights of use.

Although these arguments demonstrate that cooperative advantages theoretically did exist for Mormon settlers, their actual realization in daily life is far from being self-evident. Regardless of whether it comes to cultivating a common garden of tomatoes (Ellickson 1993) or to providing and maintaining irrigation works (Arrington et al. 1976, Worster 1985) and regardless of whether formalized rules regulating the use of a given resource exist (Benda-Beckmann et al. 2006), problems of collective action – especially incentives for free-riding or shirking – do exist in everyday life.[10] Although all community members are highly interested in profiting from irrigation infrastructure, initially there are no individual incentives to contribute to its construction and maintenance. Each individual would be better off if he or she could use the infrastructure without contributing to it. Consequently, each individual is waiting for the others to contribute to the common good.[11] In the worst case, this results in a situation where no irrigation infrastructure is provided at all. "When groups need to cooperate to achieve a collective good (such as building and running an irrigation system), strong temptations exist for participants to 'hold out and not contribute" (Ostrom 2011: 49). Consequently, there are strong incentives for free-riding which means that "those who do not purchase or pay for any of the public or collective good cannot be excluded or kept from sharing in the consumption of the good, as they can where noncollective goods are concerned" (Olson 1965: 15).

Apparently, the Mormon settlers were able to overcome these problems of collective action by using their internal governance structures. But what does this actually mean with regard to daily routines and practices? How do social interactions have to be designed in order to avoid such welfare losses? Or to put it in terms of transaction cost analysis, how can mutual monitoring and (third-party) enforcement be integrated into daily routines and practices? In this context, the perspective of eco-determination of culture suggests that Mormon pioneers should be characterized first by behavioral strategies which are functionally appropriate to sustain cooperation (and to prevent free-riding and shirking) and second by cultural preferences and value orientations which institutionally stabilize and ideologically legitimize these behavioral preferences.

In general, the daily life of a Mormon pioneer was under the influence of all kinds of (religious) rules and habits that – from a functional perspective – fostered the emergence of mutual trust and reputation as well as reciprocal monitoring and sanctioning which – in turn – are key preconditions for dealing with problems of collective action and for sustaining long-term cooperation (North 1990, Poteete et al. 2010). In this context, the "Law of Consecration and Stewardship" as well as the role of local bishops as described by Leonard Arrington et al. (1976) are of key importance.

> Briefly, the law was a prescription for transforming the highly individualistic economic order of Jacksonian America into a system characterized by economic equality, socialization of surplus incomes, freedom of enterprise, and group economic self-sufficiency. Upon the basic principle that the earth and everything on it belongs to the Lord, every person who was a member of the church at the time the system was introduced or became a member thereafter was asked to "consecrate" a deed or all his property, both real and personal, to the bishop of the church. The bishop would then grant an "inheritance" or "stewardship" to every family out of the properties so received, the amount depending on the wants and needs of the family, as determined jointly by the bishop and the prospective steward.
>
> Ibid.: 15

This quotation shows that after consecration, the Church was entitled to exercise direct control over community members and their mode of subsistence – or commercial (household) production. It was up to the local bishop to assess the amount of welfare services and stewardships – that is, temporary rights of use regarding, for example, horses, wagons, irrigation works, or a mill – that were granted to individual community members or single households. Consequently, Church members had strong incentives to live up to the expectations of the Church in order to gain lifelong group affiliation and related privileges with regard to the common good. In case they would not comply, stewardships and related rights of use could also be revoked, as in the case of laziness or in case the granted stewardships were used in selfish ways (instead of using them for the benefit of the whole community).

Apart from irrigation works, and when looking at Mormon farmland regimes, however, it could be argued that the structural moments of communalism and cooperation were not blindly applied to all tasks and challenges in the same way, but were instrumentalized in purposeful and deliberate ways. As elaborated by Arrington et al. (1976) and Ellickson (1993: 1339 ff.), the strong general preferences for cooperative behavior characterized Mormon farmland regimes only to a limited degree. Whereas families were entitled to private homesteads and gardens to cultivate their own foodstuffs, many "of the newly surveyed farms were clumped together in a 'Big Field' with a common perimeter fence, a portion of which each Big Field farmer was obliged to maintain" (Ellickson 1993: 1340). Based on these observations, it could be concluded that Mormon communities tried to avoid the disadvantages of group ownership in land as far as possible (for example, via private gardens) and – simultaneously – to reproduce a sense of communality by exploiting advantages of group ownership wherever feasible (for example, efficiencies of territorial scale in the case of fencing or irrigation).

Against the background of these fundamental peculiarities of Mormon society, the question of how monitoring and (third-party) enforcement (which can be seen as crucial preconditions for granting stewardships and related temporary rights of use) were integrated into daily routines and practices is of special interest. As elaborated by North (1990), "effective third-party enforcement is best realized by creating a set of rules that then make a variety of informal constraints effective" (ibid.: 35) – and that exactly is what the Mormon Church did: monitoring and enforcement were accomplished by hierarchic coordination (command) in combination with specific forms of labor organization (especially joint work allowing for mutual control), and religious socialization (Arrington et al. 1976). Usually, the Church laid out the general course of action– for example, that irrigation works had to be constructed and maintained in a joint effort and how much water each household was entitled to tap. On a day-to-day basis, frequent and joint efforts in constructing irrigation works or other common infrastructure allowed for a certain level of transparency and mutual control amongst the settlers (external constraints). In addition, self-constraints were stimulated by religious education and socialization as well as by the internalization of related norms, values, and behavioral codes (for example, via religious rites such as baptism, regular service attendance, or other observable and godly behavior). Free-riding, shirking, or overexploitation would not only be perceived as betraying ones' sisters or brothers in faith, but also as betraying the good Lord himself. Just in case frequent interaction (fostering mutual control and reputation building), self-constraints, and regulation by the bishop should not suffice to stabilize cooperative behavior, general conditions of daily life exposed individual community members to informal sanctioning strategies to cope with deviant behavior such as gossip and slander. In severe cases, deviators had to undergo excommunication.[12] "Where else, in the America of 1871, could a court compel reluctant members of a canal association to pay their assessment with a decision threatening that unless payment were made soon defendants would 'be cut off from the Church and their names published'" (Arrington/May 1975: 20). With this said, and from the perspective of institutional economics, the Mormon pioneers were able to profit from punishment provided by the Church whereas non-Mormon pioneers were not subject to third-party enforcement but found themselves in rather precarious and laissez faire-like conditions hampering cooperative behavior.[13]

From the perspective of eco-cultural coevolution, the physical properties of Salt Lake Valley thus selected for pioneers who were able to utilize cooperative advantages derived from division of labor and economies of scale. Cooperative behavior, however, cannot be taken for granted but highly depends on internal governance structures preventing dilemmas of collective action, especially free-riding and shirking. In the case of Mormon pioneers, these governance structures manifest themselves in religiously motivated procedures (for example, the Law of Consecration and Stewardship) and daily routines and practices which are characterized by frequent social interactions and – for that reason – foster and honor investments in good reputation. In the course of time, respective social norms and cultural preferences emerged, stabilizing and legitimizing cooperative behavioral strategies. As a result of their hierarchic organizational structure of church governance and related cultural preferences, Mormon settlers fulfilled the demanding preconditions underlying cooperative behavior and succeeded in colonizing the "American Sahara" (Arrington/ May 1975).

Summing up, and by taking tomato gardens and irrigation works for illustration, the foregoing dealt with the ecological determination of cultural preferences. Using theoretical arguments from transaction cost analysis and economic functionalism, it was argued that cultural preferences for solitary action are always established once the costs of cooperation outweigh the respective advantages. Strong cultural preferences for cooperative action, on the contrary, emerge when the benefits of cooperation compensate its costs.

Of course, cultural preferences do not directly result from the properties of physical nature, but emerge in response to the formation of appropriate behavioral strategies which – in turn – depend on the occurrence of cooperative advantages. The occurrence of advantages or disadvantages greatly affects daily routines and practices, for example, in terms of frequency, "reputation building and other relationship-specific investments and thereby alters the evolution of common norms and the degree of social ties" (Henrich et al. 2004: 46). Thus, physical environments do not only select for fit behavioral strategies and technologies, but also for functionally adequate cultural preferences stabilizing and legitimizing the adopted course of action. In the course of time, these cultural preferences become more and more institutionalized and condensed and manifest themselves as overall "cultural biases" as described by Veblen (1990), Brette (2003), and Bednar and Page (2007). Because cooperative advantages can be derived from labor division, economies of scale, and allometry which – in turn – constitute universally valid rules, it could be argued that the principles according to which cooperative advantages or disadvantages turn into cultural preferences are universally valid, too.

With this said, the second structural moment of habitat emergence, namely the cultural determination of physical nature, society, and related habitat-specific risks, will be addressed in the following chapter.

Notes

1 According to Douglas North (1990) or Gary Becker and Kevin Murphy (1992), high coordination costs especially arise in intricate exchange relations, for example, when the quality and quantity of a given good or service are difficult to assess, and which – for these reasons – depend on frequent and intense negotiation, monitoring, and enforcement efforts in order to cope with information problems as, for example, described by the principal-agent problem.

2 In contrast, socially connected technologies demand the division of labor, long-term planning, and hierarchic coordination (for example, building levees or atomic power plants).

3 Hunter-gatherer societies are another example. In contrast to the Lamalera, however, their well-being depends on covering large areas of land in order to gather, for example, mushrooms, berries, or grains and to hunt small game. In this case, solitary action presents a superior behavioral strategy, particularly because these activities are prone to free-riding and shirking. Given these conditions, and due to the fact that monitoring would be a very costly endeavor (because gathering clan members move around to cover a large area of land), solitary action presents the appropriate behavioral strategy here.

4 From our own everyday experience we know that the actual process of market exchange (for example, when buying groceries in the supermarket) is a comparably easy endeavor. However, this practical simplicity should not blind us to the fact that market transactions are complex phenomena which highly depend on institutionalization and regulation (for example, in the case of standardizing metrics or guaranteeing and enforcing property rights via third-party enforcement by the Leviathan).

5 The same principle holds true for penguins. Those living close to Antarctica tend to be bigger than their equatorial conspecifics. Whereas Antarctic penguins lose less body heat due to their comparatively smaller surface, the rather small equatorial penguins are better prepared to cope with overheating.

6 Bettencourt et al. (2007) report that by doubling the population of a city, its material infrastructure (for example, the length of electrical cables or wastewater pipes) only increases by 0.8. For mammals, the general results are that doubling the body mass of a given mammal, its metabolism rate only increases by approximately 0.75. The same holds true for their cardiovascular system. Consequently, the more body mass a mammal gains, the slower its metabolism rate and pace of life. Here, one of the most noted examples is the idea that an elephant is just a scaled-up ape which in turn is described as a scaled-up mouse (Haldane/Smith 1985).

7 Doubling the population of a city increases the rate of various interactional outcomes such as the rate of crimes, infections, or innovations by 1.2, probably because of (positive and negative) effects of social contagion due to higher social interconnectedness.

8 Of course, land regimes can also be stipulated by the Leviathan. In these cases, land regimes are commonly derived from predominant ideological values and express political interests of power, for example, in the case of the collectivized Russian agriculture or China's Great Leap Forward (Ellickson 1993:1318). However, and due to the fact that the present chapter primarily deals with the ecological determination of culture (and not with the cultural determination of nature), the case of land regimes stipulated by the Leviathan is only discussed marginally here.

9 "The case for private ownership of farms and homesteads rests on the plausible assumption that vital agricultural, construction, homemaking, and child-rearing activities entail mostly small and medium events" (Ellickson 1993: 1335).
10 "In these situations, each individual hopes to limit his or her own costs while benefiting from the contribution of others, a practice Mancur Olson (1965) referred to as 'free-riding'. The socially optimal outcome could be achieved if everybody 'cooperated'. No one is independently motivated to cooperate, however, given the predicted lack of cooperation by others. Such situations are *dilemmas* because at least one outcome yields higher returns for *all* participants, but it is not predicted that participants will achieve this outcome (Liebrand, Messick, and Wilke 1992). Social dilemmas thus involve a conflict between individual rationality and optimal outcomes for a group (Lichbach 1996; Schelling 1978; Vatn 2005)" (Poteete et al. 2010: 32).
11 In the relevant literature water is commonly conceptualized as a public good, meaning that neither excludability nor rivalry regarding its consumption exists (Ostrom 2011). Because water was a scarce resource in Salt Lake Valley, it is described as a common good, here.
12 As a result of consecration and the fact that investments in irrigation works or other public infrastructure are highly site-specific, the excommunication of a single member or family from the Mormon community would mean that all their former belongings as well as all investments in the common good would be irrecoverably lost.
13 As pointed out by North (1990), "[…] punishment is often a public good in which the community benefits but the costs are borne by a small set of individuals […]" (ibd.: 57). Therefore, it is unlikely that ordinary settlers were able to establish permanent sanctioning systems for and by themselves.

Bibliography

Abel, W. (1967). *Agrarpolitik*. Göttingen: Vandenhoeck & Ruprecht.
Arrington, L. J., & May, D. (1975). "A Different Mode of Life": Irrigation and society in nineteenth-century Utah. *Agricultural History*, 49(1), 3–20.
Arrington, L. J., May, D., & Fox, F. Y. (1976). *Building the city of God: Community & Cooperation among the Mormons*. Salt Lake City: Deseret Book.
Becker, G. S., & Murphy, K. M. (1992). The division of labor, coordination costs, and knowledge. *The Quarterly Journal of Economics*, 107(4), 1137–1160.
Bednar, J., & Page, S. (2007). Can Game(s) Theory explain culture? The emergence of cultural behavior within multiple games. *Rationality and Society*, 19(1), 65–97.
Benda-Beckmann, F. v., Benda-Beckmann, K. v., & Wiber, M. (2006). *Changing properties of property*. New York: Berghahn Books.
Bettencourt, L. M., Lobo, J., Helbing, D., Kühnert, C., & West, B. (2007). Growth, innovation, scaling, and the pace of life in cities. *Proceedings of the National Academy of Sciences*, 104(17), 7301–7306.
Beus, C., & Dunlap, R. (1994). Endorsement of agrarian ideology and adherence to agricultural paradigms. *Rural Sociology*, 59(3), 462–484.
Bowles, S., & Gintis, H. (2011). *A cooperative species*. Princeton: Princeton Univ. Press.
Bowles, S., Edwards, R., & Roosevelt, F. (2005). *Understanding capitalism: Competition, command, and change* (3rd ed.). New York: Oxford Univ. Press.
Brette, O. (2003). Thorstein Veblen's theory of institutional change: beyond technological determinism. *The European Journal of the History of Economic Thought*, 10(3), 455–477.

Coman, K. (1911). Some unsettled problems of irrigation. *American Economic Review*, 101(1), 36–48.

Dahlman, K. J. (1980). *The open field system and beyond.* Cambridge: Cambridge Univ. Press.

Dalecki, M. G., & Coughenour, C. M. (1992). Agrarianism in American Society. *Rural Sociology*, 57(1), 48–64.

Demsetz, H. (1967). Toward a theory of property rights. *The American Economic Review*, 57(2), 347–359.

Douglas, M., & Wildavsky, A. (1982). *Risk and culture: An essay on the selection of technical and environmental dangers.* Berkeley: Univ. of California Press.

Elkins, S. M., & MacKitrick, E. (1993). *The age of federalism.* New York: Oxford Univ. Press.

Ellickson, R. (1991). *Order without law: How neighbors settle disputes.* Cambridge, Mass: Harvard Univ. Press.

(1993). Property in Land. *Faculty Scholarship Series,* Paper 411. http://digitalcommons. law.yale.edu/fss_papers/411

Flinn, W. L., & Johnson, D. E. (1974). Agrarianism among Wisconsin farmers. *Rural Sociology*, 39(2), 187–204.

Foner, E. (1995). *Free soil, free labor, free men: the ideology of the Republican Party before the Civil War.* Oxford: Oxford Univ. Press.

Galilei, G. (1730 [1638]). *Mathematical discourses concerning two new sciences relating to mechanicks and local motion, in four dialogues. ... By Galileo Galilei, ... With an appendix concerning the center of gravity of solid bodies. Done into English from the Italian, by Tho. Weston, ... and now publish'd by John Weston;.* London: printed for J. Hooke.

Haldane, J. B., & Maynard Smith, J. (1985). *On being the right size and other essays.* Oxford: Oxford Univ. Press.

Hall, P. A., & Soskice, D. (Eds.) (2001). *Varieties of capitalism: The institutional foundations of comparative advantage.* Oxford: Oxford Univ. Press.

Henrich, J., Boys, R., Bowles, S., Camerer, C., Fehr, E., & Gintis, H. (Eds.) (2004). *Foundations of Human Sociality: economic experiments and ethnographic evidence from fifteen small-scale societies.* Oxford: Oxford Univ. Press.

Mosca, M. (2006). *On the Origins of the Concept of Natural Monopoly.* Unpublished manuscript, Università di Lecce Department of Economics, Working Paper No. 92/45.

North, D. C. (1990). *Institutions, institutional change, and economic performance.* Cambridge: Cambridge Univ. Press.

Olson, M. (1965). *The logic of collective action; public goods and the theory of groups.* Cambridge: Harvard Univ. Press.

Ostrom, E. (1990). *Governing the commons; the evolution of institutions for collective action* (1st ed.). Cambridge: Cambridge Univ. Press.

(2011). Reflections on "Some Unsettled Problems of Irrigation". *American Economic Review*, 101(1), 49–63.

Ostrom, E., & Walker, J. (Eds.) (2003). *Trust and reciprocity: Interdisciplinary lessons from experimental research.* New York: Russell.

Ouchi, W. G. (1980). Markets, bureaucracies, and clans. *Administrative Science Quarterly*, 25(1), 129–141.

Poteete, A. R., Janssen, M. A., & Ostrom, E. (2010). *Working together: Collective action, the commons, and multiple methods in practice.* Princeton: Princeton Univ. Press.

Schmidt-Nielsen, K. (1984). *Scaling, why is animal size so important?* Cambridge: Cambridge Univ. Press.

Schulze, E. (Ed.) (2007). *Zur Betriebsgröße in der Landwirtschaft: unter besonderer Berücksichtigung der Transformationsländer.* Leipzig: Leipziger Ökonomische Societät.

Silvestre, J. (1987). "Economies and diseconomies of scale". In J. Eatwell, M. Milgate, & P. Newman (Eds.), *The New Palgrave: A Dictionary of Economics Online.* London: Palgrave Macmillan.

Smith, A. (1776). *An inquiry into the nature and causes of the wealth of nations.* London: Strahan.

Veblen, T. (1990). *The place of science in modern civilization and other essays.* New Brunswick: Transaction Publishers[1919].

Vogel, S. (1988). *Life's devices: The physical world of animals and plants.* Princeton: Princeton Univ. Press.

West, G. B., Brown, J. H., & Enquist, B. J. (1997). A general model for the origin of allometric scaling laws in biology. *Science*, 276(5309), 122–126.

Williamson, O.E. (1967): Hierarchical control and optimum firm size. *Journal of Political Economy*, 75(2), 123–138.

Williamson, O. E. (2000). The New Institutional Economics: Taking stock, looking ahead. *Journal of Economic Literature*, 38(3), 595–613.

(2003). The economics of organization: The transaction cost approach. In M. J. Handel (Ed.), *The sociology of organizations. Classic, contemporary, and critical readings* (pp. 276–286). Thousand Oaks: Sage [1981].

Wittfogel, K. A. (1957). *Oriental despotism; a comparative study of total power.* New Haven: Yale Univ. Press.

Worster, D. (1985). *Rivers of empire: Water, aridity, and the growth of the American West.* New York: Pantheon Books.

3 The cultural determination of physical nature

The present chapter asks the questions of how and why cultural preferences, in the course of time, become increasingly detached from their functional origins, permeate more and more realms of society, and – in doing so – become hegemonic guiding heuristics coordinating both the relationships among social actors and between social actors and their physical surroundings. Whereas it was argued above that physical nature selects for suitable behavioral strategies and related cultural preferences, the present chapter claims that cultural preferences, once established, develop a life of their own and select for ideationally appropriate fits between themselves on the one side and nature, technologies, and institutions on the other.

To be more precise, and for reasons of overall societal efficiency (Bednar/ Page 2007), every eco-cultural habitat can be described by one set of hegemonic and collectively shared cultural preferences (also referred to as "cultural bias") which, from an ideal-typical perspective, oscillate between the two extreme poles of either individualistic or collectivistic value preferences and related behavioral strategies of either solitary or cooperative action. For reasons of higher cognitive efficiency, reputation building, and path-specific institutional specialization, one and the same set of cultural preferences is applied to all fields of action, meaning that cultural biases not only shape social interactions, but also affect the perception and constitution of physical nature. In other words, the approach of eco-cultural adaptation claims that the relationships among social actors and the interactions between social actors and their physical surroundings comply with similar constitutional principles.

In what follows, the underlying theoretical considerations as well as empirical illustrations are discussed in further detail. To begin with, the questions of how and why cultural preferences become increasingly detached from their initial functional origin and permeate more and more realms of society will be addressed. After that, potential benefits and limits of cultural specialization and related path-dependencies will be discussed and empirically illustrated.

3.1 The emergence of cultural preferences

In contrast to most other species, human beings can be described as social learners. They are able to outsmart slow-paced biological evolution by accumulating knowledge (for example, technological skills) and cultural preferences which are inherited from the previous generations. In evolutionary economics, this phenomenon is described as cultural inheritance (van den Bergh/Stagl 2003, Henrich/ Henrich 2007), which is quite congruent with the concepts of socialization and internalization as put forward by, for example, Norbert Elias (1939) or Peter Berger and Thomas Luckmann (1966). In this context, and as suggested by Veblen (1990) and Brette (2003), the central assumption is that human beings are socialized by predominant institutional structures and cultural preferences of their time in order to become capable of acting. With this, enhancements of human capabilities do not necessarily depend on evolutionary changes in genes, but rather on social learning and cultural inheritance.

> A good imitator [...] can simply observe and learn from the other members of his group, thereby taking advantage of the accumulated experience and wisdom of previous generations. If people mostly learn from the previous generation but occasionally make additions and improvements through their own experience, experiments, or luck, culture can become an adaptive system of learned traits that accumulate through time.
>
> Henrich/Henrich 2007: 8

Thanks to the fact that social learning via imitation is less costly than individual learning by trial and error (for example, risking intoxication when individually trying to identify edible and inedible foodstuffs), "natural selection will favor cultural learning mechanisms that allow individuals to extract adaptive information – strategies, practices, heuristics, and beliefs – from other members of their social group at lower costs than through alternative individual mechanisms" (ibid.: 10). From the perspectives of social constructivism and knowledge-sociology as presented by Berger and Luckmann (1966), social learning can thus be described as an adaptive strategy during which behavioral and cultural preferences as well as collectively shared stocks of adaptive knowledge are socially (re)produced and transformed as well as individually harnessed via socialization and internalization. In short, cultural preferences and related institutional structures are reproduced because social learning and cultural inheritance present an evolutionary superior strategy (when compared with individual learning by trial and error).

Over the generations, and via habituation and institutionalization, social learning increasingly results in culture-specific paths of habitat construction and behavioral stickiness (Berger/Luckmann 1966, Bednar/Page 2007). These cultural paths are distinct but homogeneous in themselves: ordinary day-to-day routines as well as unusual challenges are processed in similar ways within a given eco-cultural habitat (regardless of whether physical nature or society is concerned). Here again, functional arguments can be put forward to explain path-internal homogeneity and related habitat-specific cultural biases and path-dependencies. Providing socially accepted and commonly shared guiding heuristics and related practices, social learning and habituation not only relieve the individual of costly learning, but also of situative decision making in known or unknown situations (Berger/ Luckmann 1966). Conducting experiments in small-scale societies, Henrich et al.

(2004) were able to show that if "trust and commitment are important parts of one's livelihood, these sentiments may be generalized to other areas of life or evoked in situations which appear similar to everyday life" (ibid.: 47). Similarly, Natalie Henrich and Joseph Henrich (2007) point out that culturally inherited guiding heuristics are likely to get transferred to hitherto unknown challenges:

> People may culturally learn beliefs, values, and/or mental models that then act as content biases for other aspects of culture. That is, having acquired a particular idea via cultural transmission, a learner may be more likely to acquire another idea because the two "fit together" in some cognitive or psychological sense.
>
> Ibid.: 11

In order to reduce cognitive burdens and costs of learning, social actors automatically apply similar cultural preferences and behavioral strategies to distinct tasks or challenges in their physical and social environment (Giddens 1984). Thus, path-specific cultural biases become more and more coherent and homogenous in the course of time, which results in an overall increase in societal efficiency by fostering mutual understanding and expectability and consequently in a reduction of transaction costs on a daily basis. Therefore, and in the course of time, cultural preferences become increasingly detached from their initial origin (for example, the functional advantages of solitary action in the case of small events such as cultivating tomatoes as described by Ellickson (1993)), permeate more and more realms of society, and – in doing so – become hegemonic heuristics guiding social interaction. In the long run, this results in a constellation where social interaction is characterized by "intra- and inter-agent behavioral consistency" and where "rational agents choose (for rational reasons) to act culturally" (Bednar/Page 2007: 66) – even though this might result in suboptimal outcomes if taken in isolation.

Up to now, this line of reasoning – that is, highlighting the overall individual and societal advantages of acting culturally – has only been applied to relations and interactions amongst social actors in the relevant literature. In this context, the approach of eco-cultural adaptation as presented here goes one step further and suggests expanding this line of reasoning to humans' interference with nature: not only do cultural preferences shape social relationships, but they also orchestrate the social perception and constitution of physical nature. Admittedly, there are some tentative attempts to be found in the relevant literature suggesting that similar constitutional principles of social and physical nature can be empirically observed (Schwarz/Thompson 1990, Gough et al. 2008). Despite these hints, however, there is no consistent theoretical approach yet as to why such empirical coincidences should occur. To give an example, the approach of Schwarz and Thompson (1990) is grounded on ecological reports claiming that only a limited number of institutions managing ecosystems can be empirically observed. Based on these reports and in combination with the Cultural Theory (Douglas/Wildavsky 1982), Schwarz and Thompson

(1990) developed a typology describing the relationship of human beings to their natural environment. Admittedly, their fourfold typology consisting of culturally biased ways of life and related myths of nature mostly suffices to describe empirically observable elective affinities between cultural preferences and related perceptions of physical nature. However, the authors fail to present satisfying theoretical arguments explaining why specific cultural patterns and myths of nature should coincide. Here, the approach of eco-cultural adaptations claims that a limited number of observable elective affinities between cultural preferences and societal ways of constituting physical nature can be theoretically explained from a coevolutionary perspective.

Taking seriously the efficiency-argument as developed above, it can first be argued that it is cognitively more efficient to transfer one and the same set of cultural patterns (and related behavioral strategies) to the coordination of human-nature relationships rather than switching between object-specific heuristics. Empirical evidence from environmental psychology supports this line of argumentation. Based on the norm-activation-model as developed by Shalom Schwartz (1970), it was possible to show that altruistic and empathic individuals (as opposed to social actors with strong individualistic and competitive value orientations) anticipate how their behavior would possibly affect the physical environment and, in turn, how these environmental impacts would affect their fellow men (Schwartz 1994, Stern et al. 1995, Stern 2000). Consequently, altruists try to avoid environmentally harmful behavior whereas individualists do not really care about negative impacts regarding their physical and social environment.

Second, constituting physical nature and society along similar lines does not only present a superior cognitive strategy, but can also be advantageous with regard to daily routines and social relationships, especially in view of high returns to relationship-specific investments and reputation building (first and foremost in the case of high cooperative advantages). The more overall social welfare depends on cooperation and frequent interaction (for example, in order to provide irrigation works or in the case of traditional whale hunting), the more important relationship-specific investments and reputation building become (Henrich et al. 2004). Refraining from environmental behavior that would harm fellow men thus becomes a means of reputation building. Take, for example, two neighbors who are bound to a place by high site-specific investments and are aware they will have to get along well with each other for a lifetime. If high benefits to long-term cooperation and reputation building exist, it is not only wise to treat one's fellow men well, but also the physical environment they live in and depend on.[1] By implication, this also means that low paybacks to reputation building and cooperation result in less frequent interaction, which, in turn, should become observable in comparably less cautious environmental behavior.

Finally, and at an advanced stage of habitat formation, similar patterns regarding the constitution of physical nature and society can also be explained by path-specific institutional specialization and socialization: institutional properties that have emerged to cope with non-excludability and related problems of collective action should easily be transferable to problems featuring similar structural properties in the realms of social welfare or environmental regulation. By contrast, it can be assumed that eco-cultural habitats featuring an institutional environment that has emerged in order to support solitary action and coordination by means of (market) competition will rather utilize market incentives (as opposed to centralized planning and hierarchic coordination) when it comes to shaping the interference with physical nature (Schwarz/Thompson 1990).

Taken together, and recalling the considerations about the ecological determination of cultural preferences discussed in Chapter 2, the conjunction of physical environments, behavioral strategies, and cultural preferences governing the constitution of physical nature and society can thus be further specified: social groups populating eco-cultural habitats primarily specialized in cooperative action do not distinguish between social cohesion, non-kin solidarity, and environmental sustainability and – for that reason – cautiously treat their physical environment like fellow men. In contrast, individuals inhabiting eco-cultural habitats mainly specialized in establishing institutional circumstances allowing for solitary action and competitive behavior are expected to instrumentalize both their fellow men and physical nature for the sake of short-term maximization of personal utility. Fellow men are assigned to the non-human physical environment and treated alike (Luckmann 1970).

Admittedly, these are strong claims. However, if these are valid it should be possible to, for example, reveal parallels between social welfare and environmental regulation on the level of nation states within distinct eco-cultural paths. Without going into detail here, empirical evidence from the relevant literature about distinct production and welfare regimes and related environmental regulations is presented in Chapter 6 (Esping-Andersen 1990, Hall/Soskice 2001, Dryzek et al. 2003, Gough et al. 2008, Schröder 2009, Gough/Meadowcroft 2011). Here, it can actually be observed that coordinated welfare and production regimes coincide with strong concerns for environmental sustainability and cooperative technologies whereas ecological issues, as well as issues of social welfare, are of comparably minor importance in liberal regimes.

Summing up, this section has addressed the questions of how and why cultural preferences, in the course of time, become increasingly detached from their initial functional origin, permeate more and more realms of society, and – in doing so – become hegemonic guiding heuristics coordinating the constitution of physical nature and society in analogous ways. To complete the theoretical argument describing and explaining the emergence of eco-cultural habitats and related cultural path-dependencies, the crucial question of how cultural biases – once established – in their turn affect the emergence and reproduction of eco-cultural habitats will be considered next.

3.2 Cultural stickiness and its ambivalent effects on eco-cultural habitats

The present subchapter problematizes the question of how the (re)production and stabilization of eco-cultural habitats are affected by the self-continuation of culture and the emergence of hegemonic cultural biases. With this, and in the wording of Brette (2003), we are now looking at an advanced stage of habitat formation where the prevailing institutional structure starts to exert:

> a selective role among the instinctive dispositions of the population. [...] All in all, the cultural scheme not only selects some instincts – and counter-selects some others – but it denatures them as to make them coherent with itself. [...] This simply means that, once established, the institutional system tends to become relatively autonomous and to impose itself on instincts, individual habits and new institutions so as to make them consistent with itself.
>
> Ibid.: 464

In other words, cultural preferences have become more or less detached from their original functional source and finally permeate all realms of social action. As a result of habituation and institutionalization, cultural preferences and related behavioral strategies have turned into overarching cognitive and emotional dispositions providing social actors with normatively charged guiding principles orchestrating the constitution of physical nature and society. Now, hegemonic cultural preferences, in turn, select for ideational fits between themselves on the one side and nature, technologies, and institutions on the other. Thereby, functional necessities or the existence of cooperative advantages as discussed above are of only minor importance or even completely neglected. Thus, and using Anthony Giddens'(1984) terminology, cultural preferences have become part of what is called "practical consciousness" and are – in most cases unconsciously! – applied to any kind of socio-ecological task or challenge within the respective eco-cultural habitat, be it environmental issues such as climate change or social issues such as dismissal protection or health insurance. Therefore, it is assumed that habitat-specific cultural path-dependencies establish themselves in the course of time.

With this said, a sheer functionalistic perspective as developed in Chapter 2 no longer suffices. Therefore, the complementary analytical perspective of social constructivism (Berger/Luckmann 1966) seems to present an appropriate tool to observe how culturally biased ways of life and related ideational and normative beliefs influence the adaptive capacities of eco-cultural habitats. Social constructivism – in contrast to economic functionalism – highlights the important fact that matches between nature, technologies, institutions, and culture are not just predetermined by the occurrence of cooperative advantages or disadvantages as argued above, but – once hegemonic cultural preferences have become established – increasingly depend on the culturally biased construction of social and physical nature. In this context, the constructivists' paradigm as encapsulated by William Thomas (1928) is of particular importance: "If men define situations as real, they are real in their consequences" (ibid.: 571).

Trying to transfer the Thomas theorem to the formation of eco-cultural habitats, the central question is how cultural persistence (and related forms of cultural selectivity) influences the constitution of physical nature and society and how these constitutional processes, in turn, shape the matches between physical nature, technologies, and institutions. In other words, what situations and world views do inhabitants of distinct eco-cultural habitats define as being real? What are the culturally preferred physical and social environments, technologies, and institutions and how are they socially constructed? And finally: how do these culturally biased world views and preferences affect the formation and reproduction of eco-cultural habitats? In order to answer these questions, path-specific world views and related principles regarding the constitution of physical nature and society must be given more attention, an endeavor for which the Cultural Theory promises to bring to light further insights (Douglas/Wildavsky 1982).

From the perspective of the Cultural Theory, it can be argued that any cultural path is characterized by a distinct (i.e. culturally biased) world view which defines a normatively preferred way of life. Thereby, each of these normatively preferred ways of life is accompanied by specific conceptions of ecological and societal risks which are believed to endanger further habitat existence by threatening social and environmental stability.[2] In order to secure environmental and social stability, each way of life defines specific (and also culturally biased) mechanisms of defense. Depending on the specific properties of these selected risks and defense mechanisms, each world view stipulates normative rules and ideological superstructures regulating and legitimizing social relationships as well as humans' interference with physical nature (Douglas/Wildavsky 1982, Schwarz/Thompson 1990, Mamadouh 1999, Verweij et al. 2006). As described by the Thomas theorem and emphasized by the Cultural Theory, these culturally biased world views thus determine what is defined as relevant and real, what men act upon (or do not act upon), and – as a result – what becomes efficacious social reality. With regard to the formation of eco-cultural habitats and their overall texture as illustrated in Figure 1.1, this means that culturally biased world views try to make physical nature, technologies, and institutions consistent with themselves (Veblen 1990, Brette 2003, Bednar/Page 2007).

Given these considerations, physical nature and society as well as technologies or institutions do not exist independent from human conceptions, but are socially constructed and moldable. Similarly, concepts such as "transaction cost-economizing governance structures" or "cooperative advantages" as introduced above do not exist in an objective or culturally unbiased way (as implicitly claimed hitherto), but are subject to cultural preferences, too. In other words, culturally biased ways of life also define what is perceived as being advantageous or costly (Henrich et al. 2004, Henrich/Henrich 2007). What might appear as being costly from the perspective of economic functionalism is not necessarily perceived as being costly culturally. Whereas economic functionalism would claim that task-specific governance structures should always be applied to reduce the costs associated with the coordination of cooperation (for example, in the case of cultivating tomatoes, producing and distributing toothbrushes, or providing flood protection), the culturalistic perspective reveals that one and the same cultural bias – for reasons discussed above – is applied to all tasks, even though this might result in suboptimal outcomes if taken in isolation (Berger/Luckmann 1966, Bednar/Page 2007). Thus, the constructivist perspective as presented here seriously questions the functionalistic considerations presented above, at least at a first glance. At a second glance, however, addressing the complementary (and sometimes conflicting) dynamic between functional necessities and cultural preferences is exactly what uniquely characterizes the theoretic approach of eco-cultural coevolution (as opposed to sheer realism or constructivism as sketched out above). The formation of eco-cultural habitats is neither ecologically nor culturally determined. It is the very dynamic of these two complementary forces which actually characterizes the emergence of eco-cultural paths of adaptation (Veblen 1990, Brette 2003).

Next and as illustrated in Table 3.1, an attempt is made to sketch a theoretically motivated typology of culturally biased ways of life that captures the complementary (and potentially tension-filled) structural moments of functional necessities and cultural preferences and their (ambiguous) effects on the formation of eco-cultural habitats. To begin with, and terminologically inspired by the relevant literature about varieties of capitalism as introduced by Peter Hall and David Soskice (2001), it is claimed that cultural paths of eco-cultural adaptation oscillate between the two extreme poles of either liberal or coordinated paths. As elaborated in Chapter 2, liberal cultural paths initially originate from cooperative disadvantages which foster solitary action and entail the emergence of cultural preferences for strong individualism, individual autonomy and freedom. Accordingly, the central feature characterizing the liberal institutional environment is to enable solitary action and the coordination of social relationships via competitive means. In the same way, social actors inhabiting liberal habitats prefer socially unconnected technologies which can be applied individually and spontaneously. Solidarity is restricted to kin, clan, or ethnicity.

Coordinated cultural paths, in contrast, originate from cooperative advantages encouraging cooperative behavioral strategies and reputation building which, in turn, entail the emergence of rather collectivistic cultural preferences. Consequently, the coordinated path is accompanied by institutional structures fostering the coordination of social relationships via hierarchic regulation and favors socially connected technologies which can only be collectively applied via hierarchic coordination. Analogously to socially connected technologies, solidarity is not restricted to kin or clan but also includes strangers.

Furthermore, and due to different returns to foresight, it can be assumed that liberal and coordinated paths differ with regard to time preferences. Because cooperative advantages can only be utilized by cooperative behavior which, in turn, highly depends on reputation building (meaning that social actors have to anticipate how contemporary actions will affect social relationships in future times), it can be expected that coordinated cultural paths are characterized by long-term planning and comparatively high time preferences. High payoffs to solitary action, in contrast, diminish the importance of foresight and reputation building and, for these reasons, should be accompanied by comparably low time preferences (Elias 1939). As discussed above, different returns to foresight and reputation building – in the long run – are not only assumed to affect the extent of solidarity and social welfare amongst social actors, but also to shape human interference with nature. In this context and also taking into account different time

Table 3.1 Path-specific specialization and related strengths and vulnerabilities due to cultural selectivity and persistency

Cooperative advantages	Low	High
and	Adaptation individually and spontaneously achievable, low paybacks to reputation building.	Adaptation requires collective effort and long-term perspective, high paybacks to reputation building.
type of behavioral strategy	→ Solitary action	→ Collective action
Time preference	Short	Long
Compatible cultural preferences	Individual freedom and autonomy, solidarity limited to kin or ethnicity.	Collectivism, solidarity among strangers.

Homogenous culturally biased ways of life emerge in the course of time, especially mediated by social learning and cultural inheritance and due to higher cognitive and societal efficiency (as opposed to individual learning, situational decision making, and juggling with many object-specific decisional heuristics).

Culturally biased way of life	Liberal cultural bias	Coordinated cultural bias
Institutional environment guiding the constitution of social and physical nature	Ranges from laissez faire/self-subsistence to market-coordination.	Ranges from clan-like governance structures to hierarchic coordination
Preferred technologies	Socially unconnected, individually and spontaneously applicable.	Socially connected, collectively applicable, based on long-term planning and centralized regulation.
Perception of social and physical nature	Perceived as resilient means to the end of individual utility maximization in a short-term perspective.	Perceived as fragile ends in their own rights, valorization of social and physical nature take long-term social welfare and ecological sustainability into account, strong preferences for the public good.
Specialized in …	Challenges and technologies concerning individual actors and demanding for decentralized, spontaneous, and unbureaucratic actions and decision making.	Challenges and technologies which involve large social collectives, embrace time and space, and depend largely on hierarchic coordination and long-term planning.
Vulnerable to …	Challenges which go beyond individual capabilities and demand for long-term planning and hierarchic coordination.	Challenges which demand for immediate response and short-term flexibility.

Source: compiled by the author.

preferences, it is expected that the valorization of physical nature rather follows individual short-term interests within liberally biased paths. The physical environment is treated as an individual means rather than as a collectively relevant end in its own rights. Of course, physical nature is instrumentalized for individual means in coordinated paths, too. However, and in contrast to the liberal path, valorization of nature does not follow laissez faire-like conditions, but is generally guided by strong preferences for the collective good, high levels of social welfare, and long-term ecological sustainability (Dryzek et al. 2003, Gough et al. 2008).

Against this background and seen from the perspective of social constructivism, the central argument is that eco-cultural habitats operate like this: socio-ecological tasks and challenges such as flooding, drought, or unemployment are filtered and redefined in such a way as to comply with the predominant views of society and nature. In this way, they become processable by habitat-specific technologies and institutions. Thus, culturally biased world views reduce overall complexity by exercising a specific form of selectivity. Based on this path-specific selectivity, (culturally biased) social action and related ways of constituting eco-cultural realities take place (Thomas 1928, Berger/Luckmann 1966). Here, Cultural Theory (Douglas/Wildavsky 1982) can be harnessed to further specify the hidden codes steering cultural selectivity within liberal and coordinated eco-cultural paths of adaptation: the beloved principles regulating social relationships and human interference with nature provide suitable solutions to propagated risks. With this said, the following typology of ideal-typical prototypes of eco-cultural habitats can be put forward now.

In the case of liberal eco-cultural paths of adaptation, and depending on the overall level of formalization and bureaucratization, the central principles coordinating social action are assumed to range from sheer laissez faire-like conditions (as in the case of trappers or pioneers left to themselves) to anonymized exchange and third-party enforcement, for example, via competitive mechanisms such as (spot-)markets as in the case of liberal production and welfare regimes (Esping-Andersen 1990, Hall/Soskice 2001). Accordingly, potential risks threatening further habitat existence range from any kind of restrictions concerning individual freedom and autonomy to all kinds of limitations of competitive exchange relations.

The central organizational principle that characterizes social action within coordinated adaptational paths, in contrast, is hierarchic coordination and regulation. Depending on the overall level of formalization and bureaucratization, hierarchic coordination ranges from informal clan-like governance structures to highly formalized coordination by the Leviathan, for example, in the case of social-democratic production and welfare regimes (Esping-Andersen 1990, Hall/Soskice 2001) or ancient despotic hydraulic societies (Wittfogel 1957). Consequently, all kinds of unpredictable conditions (for example, volatile social and physical environments challenging hierarchic control, planning, and related bureaucratic routines) are perceived as fundamental risks threatening further habitat existence.

From the perspective of an external observer, and due to these different types of path-specific cultural selectivity and related constructions of reality, both types of path have a Janus-faced character regarding their capabilities of eco-cultural adaptation. In general, the advantage of path-specific cultural selectivity can be seen in the fact that each habitat becomes highly specialized with regard to a given set of socio-ecological challenges. From this perspective, the liberal eco-cultural habitat is especially predestined to deal with socio-ecological challenges which concern individual actors (or small collectives such as local populations or single firms) and demand for decentralized, spontaneous, and unbureaucratic actions, decisions, and related technologies within short- or medium-term planning horizons. The coordinated eco-cultural habitat, in contrast, is destined to deal with socio-ecological challenges and technologies which involve large social collectives (or even future generations), embrace time and space, and – for these reasons – largely depend on hierarchic coordination and long-term planning. With this, it is also claimed that coordinated habitats are primarily specialized in the application of connected technologies as well as in the provision of non-excludable goods, whereas liberal habitats are primarily specialized in the application of socially unconnected technologies as well as in the provision of excludable goods.

At the same time, however, an external perspective also reveals that path-specific specialization and related constructions of reality are bought at the price of particular vulnerabilities. The liberal habitat is especially vulnerable to socio-ecological challenges which go beyond individual capabilities and demand for long-term planning and hierarchic coordination. Further, it can be assumed that non-excludable goods such as flood protection, public security, or social welfare are only sparsely provided (if at all), which, in turn, carries new vulnerabilities in its own right, for example, regarding the effects of social inequality (Wilkinson and Pickett 2009). Coordinated habitats, on the contrary, are particularly vulnerable to socio-ecological challenges which demand immediate response and short-term flexibility. Moreover, and due to path-specific specialization, coordinated regimes arguably tend to handicap themselves by exuberant bureaucratic procedures and administrative measures (Fourcade-Gourinchas/Babb 2002).

Thus, path-specific selectivity and related forms of specialization are not only beneficial, but also go hand in hand with specific forms of vulnerabilities and mismatches. In general, and from an external perspective, mismatches occur whenever functional necessities and cultural preferences contradict each other. Because cultural biases exert selective power over the constitution of physical nature and society, functional necessities become molded by the predominant cultural bias. To give some examples, this means that liberal habitats also try to provide non-excludable goods such as pollution-free air. However, and in accordance with their liberal bias, this is not directly achieved by hierarchic coordination or stipulations, but rather by sharpening market incentives, for example, for emission certificates, otherwise environmental protection would not be processable within the liberal habitat. Analogously, the provision and distribution of toothbrushes or apples normally is not coordinated by the Leviathan (for example, via ration coupons) but via market mechanisms in coordinated habitats.[3] However, potentially negative effects of market coordination are mitigated by means of hierarchic coordination (for example, via strong social welfare arrangements and environmental regulation). In short: if functional necessities and cultural preferences are not compatible, functional necessities may be culturally overridden. It is therefore an empirical question of whether functional necessities are totally ignored or taken into account sufficiently.

In this context, Bednar and Page (2007) point out that cultural behavior admittedly results in overall societal efficiency in general, but often eventuates in suboptimal adaptational measures when taken in isolation (ibid: 65). However, and as long as the preferred way of life can be ecologically sustained, mismatches are of no further practical consequence. Besides, and as will become apparent in Chapter 6, liberal and coordinated habitats have developed ideological superstructures legitimizing (or even idealizing) minor mismatches. To give an example, social myths such as "rugged individualism", the "up-by-the-bootstraps philosophy", or the "rags to riches" tale emerged in the liberal habitat in order to cope with high degrees of social inequality (which can be interpreted as an unintended byproduct of cultural preferences cherishing solitary action).

Only when culturally biased ways of life become ecologically unsustainable are habitats critically endangered. To give an example, this might be the case when ecological limits or functional necessities are totally ignored, blocked out, or refused acceptance on cultural or ideological grounds, as, for example, in the case of collectivized Russian agriculture or China's Great Leap Forward, which resulted in severe famine and a huge death toll (Merl 1985, 1990). If such constellations occur, the approach of eco-cultural adaptation predicts increasing pressure to adapt, meaning that more appropriate matches between physical nature and ideational preferences are selected for, which, in turn, results in habitat-transformation. Otherwise, and as described by Albert Hirschman (1970) and Jared Diamond (2005), either exit (for example, migrating from one eco-cultural habitat into another) or collapse are likely scenarios.

Now, after the second constituent of eco-cultural habitat formation – namely the self-continuation of cultural patterns, cultural selectivity, and its influence on adaptational capacities – has been discussed, three empirical examples illustrating how culturally biased ways of life can affect the emergence and (re)production of eco-cultural habitats will be presented. In line with the empirical examples discussed in Chapter 2, the following illustrations again focus on agricultural land regimes: the collectivization of agriculture in the former Soviet Union, the Dust Bowl in the North American Great Plains, and overregulation in Mormon settlements. These cases represent consciously chosen examples for mismatches illustrating the mechanisms of cultural biases. As Chapter 6 will deal with the historical reconstruction of habitat emergence as well as with the question of how culturally biased ways of life affect adaptational capabilities in greater detail, the following historical examples are only briefly presented.

The topic of shifting agricultural land regimes characterizing the (former) Soviet Union represents an intricate matter of research (Merl 1985, Conquest 1986, Merl 1990). Regarding the question of how culturally biased world views and related ideological beliefs affect the (re)production of eco-cultural habitats, the collectivization of agriculture as enforced by Stalin between 1928 and 1940 is of special interest: it perfectly lends itself to a demonstration of how cultural biases and related ideological convictions can affect hitherto stable matches between social and physical nature. One of the central goals of agricultural collectivization was to increase agricultural productivity which, in turn, was seen as a crucial precondition for urbanization and industrialization. In accordance with the socialist conviction that, first and foremost, the means of production (for example, land) should be socialized and not be possessed individually, large-scale collective farms were believed to help raise agricultural productivity (and to even out social inequality).

As elaborated by Stephan Merl (1999), different property regimes regulating the use of farmland coexisted before the collectivization. No matter whether farmland was formally privately or collectively owned, strong customary laws generally had resulted in quasi private property in land and individual farmers were able to make a living from subsistence agriculture. With this, and as elaborated in Chapter 2, it could be argued that agricultural production regimes before collectivization were effectively characterized by what looked like individual ownership in land and related advantages (Ellickson 1993). In this context, it was argued that individual ownership in land simplifies monitoring and provides incentive structures ensuring that the responsibilities for cultivation and monitoring rest with those who have strong individual incentives to diligently fulfill these tasks. Thus, agricultural productivity is comparably high.

Consequently, it does not surprise that the collectivization of agriculture did not boost agricultural productivity (as expected by the socialist planners), but rather increased incentives for laziness, free-riding and shirking. What was boosted were the costs associated with coordinating cooperation and common property in land (as Ellickson would have predicted). To give an example, collectivized farmers were obliged to deliver their harvest to the state. In reaction, and due to the lack of individual incentives, it is reported that some farmers stopped farming or even burned the harvest. Most farmers would appropriate considerable parts of the harvest for private use for nutrition or to trade with. Despite the fact that brutal attempts were made to prevent such practices (for example, by arresting those involved in gleaning), the state did not succeed in nationwide monitoring and enforcement. Agricultural productivity sharply declined and approximately 10 million people starved (Merl 1985, 1990).

On the bottom line, the overall socio-economic situation deteriorated to such an extent that individual farmland had to be revitalized and even subsidized by the state in order to secure the workers' survival in the 1930s (Merl 1999). From the 1950s onwards, individual farmland was permanently granted. Thus, two different production regimes coexisted from thereon: large-scale collective farms and individual farmland (primarily used for self-supply, but also for trade). Although individual farmland and gardening covered only a negligible portion of the farmland available nationwide, a substantial amount of all agricultural goods was produced on this small share of acreage (Abel 1951: 214, Schulze 2002).

While forced collectivization was in full swing in the Soviet Union, farmers on the other side of the northern hemisphere suffered from severe hazards in the form of sandstorms (Worster 1979, Cunfer 2005, Montgomery 2007). Regarding the work of Ellickson (1993), it could be argued that the settlers who were the first to break the ground in the North American Great Plains around 1850 were culturally well equipped for farming. Their behavioral strategies were primarily shaped by the Jeffersonian ideal of yeoman farmers as embodied by the Homestead Acts fostering individualistic behavior and solitary action in order to obtain individual property in land – a behavioral strategy which is generally well suited to coordinate agricultural production at comparably low agency costs. In the Great Plains, however, competitive and individualistic behavior finally resulted in severe socio-ecological crises referred to as the Dust Bowl (Worster 1979).

Based on the work of Worster (1979) and David Montgomery (2007), it can be argued that the Dust Bowl embodies a crucial maladjustment between the ecological properties of the North American semiarid plains and farming practices primarily driven by the "ecological values taught by the capitalist ethos" (Worster 1979: 6):

> It [the Dust Bowl, author's note] cannot be blamed on illiteracy or overpopulation or social disorder. It came about because the culture was operating in precisely the way it was supposed to. [...] Some environmental catastrophes are nature's work, others are the slowly accumulating effects of ignorance or poverty. The Dust Bowl, in contrast, was the inevitable outcome of a culture that deliberately, self-consciously, set itself the task of dominating and exploiting the land for all it was worth.
>
> Ibid.: 4

Apparently, capitalist culture and the properties of the North American semiarid plains – first and foremost low precipitation, strong winds, and fertile (but fine-grained and lightweight) loess – do not go together when paired under laissez faire-like competitive conditions. Driven by the strong desire to find a home and live an independent life, and supported by technological innovations like the steel plow and tractors, settlers – everyone for themselves – began breaking and cultivating the Great Plains. Settling the Great Plains was also promoted by the "rain follows the plow" theory, according to which homesteading would make semiarid regions more humid (which, as we all know, did not happen, Worster 1979). However, the sheer fact that such a myth came up at all and that people believed in it and consciously voted for subsistence and cash cropping in semiarid regions demonstrates the strength of their wishful thinking. By turning the ground, settlers destroyed the grass cover and the rootage that had formerly protected the finely grained and lightweight soil from erosion by wind and water. Now, the bare soil lay there unprotected against erosion for most of the year. In consequence, and combined with a series of severe droughts during the 1930s, more and more sandstorms occurred. Living and farming conditions in the Great Plains rapidly deteriorated.[4]

In Congress, the amount of erosion triggered a discourse about whether "the rapid pace of soil destruction threatened to undermine civilization" (Montgomery 2007: 148) and resulted in the foundation of a new Soil Conservation Service (ibid.: 152). "The new task involved safeguarding, with public power, privately owned and privately worked land. It demanded, according to the planners, fresh thinking about property rights of the individual where they threatened the community's welfare" (Worster 1979: 186). Without going deeper into this matter here, this can be interpreted as an attempt at habitat stabilization and reconstruction by replacing a land-use regime ruled by unregulated individual short-term interests by one guided by hierarchic coordination and long-term planning (for example, regarding ecological sustainability). However, the principles of hierarchic coordination were in stark contrast to the overly powerful cultural preferences "taught by the capitalist ethos" (Worster 1979: 6). The attempt at habitat reconstruction via the implementation of novel institutional structures more or less failed. Nonetheless and in accordance with the predominant cultural bias and related coordinative competitive principles, farmers – at least those who had not voted for exit – individually adapted to the difficult growing conditions in the semiarid plains, especially via novel (and capital-intensive) technologies and farming practices – for example, irrigation, dry farming, zero tillage, the planting of cover crops and hedges to break the winds, or contour plowing to prevent water erosion.

In Richfield, Utah, Mormon pioneers struggled with comparable structural problems in the second half of the nineteenth century. In the same way as settlers in the Great Plains transferred individualistic preferences in a laissez faire-like way to a rather fragile ecosystem, Mormon pioneers applied their collectivistic cultural bias and hierarchic coordination not only to the problem of irrigation, but generally to all realms of social life. To give an example, it can be shown how the principle of equality – a fundamental element of Mormon collectivism – prevented the formation of prices and – as an unintended byproduct – resulted in high coordination and negotiation costs as well as a high level of dissatisfaction which – as in the case of Richfield – finally resulted in the local dissolution of the United Order (Arrington et al. 1976).

"The principle of equality required the adjustment of wages, piece rates, and prices, so that farmers, unskilled laborers, and mechanics would receive approximately the same credits and work approximately the same number of hours" (ibid.: 181). Community members were not paid in cash, but their piece rates and working hours were centrally registered in the Order's records. In the local stores, where the community's surplus was managed, community members were able to exchange these credits for clothing or food.

This system was accompanied by several serious problems. To begin with, and due to equal pay and the absence of personal incentives, daily work was characterized by indolence and indifference. In response, dissenters were urged to be diligent (or were threatened with expulsion) by the local Church authorities. In other words, the principle of equality and respective peculiarities such as having equal credits instead of task-specific wages provided strong incentives for free-riding and shirking which, in turn, resulted in increasing monitoring, negotiation, and enforcement efforts. In addition, accounting proved to be a rather slow and tardy endeavor and individual settlers as well as Church authorities had only vague ideas about their current financial balance between credits and debts, a situation which blurred the fact that more was consumed than was produced. "Thus, it was impossible for the secretary to keep a current record of individual labor credits, and it was impossible for an individual to find out where he stood, or for the board to find out if he was not already overdrawn" (Arrington et al. 1976: 194). As one can imagine, the overall supply situation deteriorated and the dissatisfaction among the community members steadily increased. Those who threatened to seek paid labor outside the community – for example, to earn some extra money to buy shoes or food – were cut off from their supplies until they conformed to the Order's rules (ibid.: 195). Eventually, even the semi-monthly community meetings giving "disgruntled members an opportunity to release a good deal of pent-up emotions" (ibid.: 187 ff.) proved to be ineffective and the Order finally broke down. It can be concluded that the Richfield community collapsed due to exuberant costs of coordination caused by overregulation which, in turn, is symptomatic of a community trying to organize all realms of social life along similar ideational and coordinative lines.[5]

With these examples illustrating the potentially negative consequences of cultural selectivity and persistency, the theoretical approach of eco-cultural adaptation has now come full circle. To sum up, this chapter has illustrated how and why cultural preferences, in the course of time, become increasingly detached from their initial functional origin, permeate more and more realms of society, and – in doing so – become hegemonic guiding heuristics coordinating both the relationship among social actors and between social actors and their physical surroundings. Further, habitat-specific culturally biased ways of life and related modes of constructing social reality have been discussed and ideal-typically contrasted. Thereby, the Janus-faced character of cultural persistency and selectivity was of special interest: path-specific specialization and distinct forms of adaptational capabilities are bought at the price of (more or less severe) maladjustments and particular vulnerabilities. With this, the particular character of eco-cultural coevolution was finally touched: in contrast to unidirectional approaches dealing with the relationship between human beings and their physical surroundings, the complementary dynamic between functional necessities and cultural preferences is exactly what characterizes both the theoretic approach of eco-cultural coevolution as well as empirical evidence. The formation of eco-cultural habitats is neither ecologically nor culturally determined. It is the very dynamic of these two complementary and sometimes conflicting forces which actually characterizes eco-cultural coevolution and the emergence of distinct adaptational paths.

Notes

1 In contrast to the authors from environmental psychology cited above, the perspective of eco-cultural coevolution argues that individuals do not necessarily anticipate the environmental consequences of their behavior for fellow men because of their altruistic value orientation (at least in the beginning of eco-cultural coevolution), but that they have

become altruists (via habituation, institutionalization, and path-specific socialization) because they have learned that frequent interaction and related investments in reputation building work more smoothly when not interfering with their neighbors' physical or social environments in harmful ways.

2 In this context, it is important to have in mind that habitat-specific concepts of social and environmental stability are culturally biased, too.

3 Admittedly, there are historic and current examples showing that the provision and distribution of specific goods was, and still is, coordinated by centralized regulation and ration coupons (for example, in the former Soviet Union or in contemporary Cuba). Here, the perspective of eco-cultural adaptation would argue that the predominant collectivistic cultural bias totally ignores functional necessities and is transferred to realms of social action which should better be organized via less costly governance structures (for example, (spot-)market exchange). Thus, the emergence of black markets does not surprise but can be interpreted as an attempt to establish less costly exchange relations.

4 Some farmers, sometimes overnight, lost all their fertile topsoil. Livestock and draft animals died of thirst, from inhaled dust and caked lungs, or just starved because the pastures were covered with dust. Some farmers committed suicide, others tried to find salvation in using sedatives, and approximately three million-plus farmers opted for exit and left the Great Plains (Worster 1979, Montgomery 2007).

5 Here, the question comes to mind why hierarchic coordination in Richfield did not work as well as in the case of irrigation works in Salt Lake Valley, as discussed in subchapter 2.2. According to Arrington et al. (1976), a tentative answer could be that the Mormon community in Richfield grew too rapidly and that those newly arrived in the community had insufficient assets to contribute to the common good. Thus, the available assets were shared by more and more people over the course of time. Finally, limits of capacity were exceeded and diseconomies of scale occurred.

Bibliography

Abel, W. (1951). *Agrarpolitik*. Göttingen: Vandenhoeck & Ruprecht.

Arrington, L. J., May, D., & Fox, F. Y. (1976). *Building the city of God: Community & Cooperation Among the Mormons*. Salt Lake City: Desert Book.

Bednar, J., & Page, S. (2007). Can Game(s) Theory explain culture? The emergence of cultural behavior within multiple games. *Rationality and Society*, 19(1), 65–97.

Berger, P., & Luckmann, T. (1966). *The social construction of reality; a treatise in the sociology of knowledge* (1st ed.). Garden City: Doubleday.

Brette, O. (2003). Thorstein Veblen's theory of institutional change: beyond technological determinism. *The European Journal of the History of Economic Thought*, 10(3), 455–477.

Conquest, R. (1986). *The harvest of sorrow: Soviet collectivization and the terror-famine*. New York: Oxford Univ. Press.

Cunfer, G. (2005). *On the Great Plains: Agriculture and environment* (1st ed.). College Station: Texas A & M Univ. Press.

Diamond, J. M. (2005). *Collapse: How societies choose to fail or succeed*. New York: Penguin.

Douglas, M., & Wildavsky, A. (1982). *Risk and culture: An essay on the selection of technical and environmental dangers*. Berkeley: Univ. of California Press.

Dryzek, J. S., Downs, D., Hernes, H.-K., & Schlossberg, D. (Eds.) (2003). *Green states and social movements: Environmentalism in the United States, United Kingdom, Germany, and Norway*. Oxford, New York: Oxford Univ. Press.

Elias, N. (1939). *Über den Prozeß der Zivilisation; soziogenetische und psychogenetische Untersuchungen.* Basel: Verl. Haus zum Falken.

Ellickson, R. (1993). Property in Land. *Faculty Scholarship Series,* Paper 411. http://digitalcommons.law.yale.edu/fss_papers/411

Esping-Andersen, G. (1990). *The three worlds of welfare capitalism.* Cambridge: Polity.

Fourcade-Gourinchas, M., & Babb, S. L. (2002). The rebirth of the liberal creed: Paths to neoliberalism in four countries. *American Journal of Sociology,* 108(3), 533–579.

Giddens, A. (1984). *The constitution of society: Introduction of the theory of structuration.* Berkeley: Univ. of California Press.

Gough, I. et al. (2008). JESP symposium: Climate change and social policy. *Journal of European Social Policy,* 18(4), 325–344.

Gough, I., & Meadowcroft, J. (2011). Decarbonizing the Welfare State. In J. S. Dryzek, R. B. Norgaard, & D. Schlosberg (Eds.), *The Oxford handbook of climate change and society* (pp. 490–503). Oxford: Oxford Univ. Press.

Hall, P. A., & Soskice, D. (Eds.) (2001). *Varieties of capitalism: The institutional foundations of comparative advantage.* Oxford: Oxford Univ. Press.

Henrich, J., Boys, R., Bowles, S., Camerer, C., Fehr, E., & Gintis, H. (Ed.) (2004). *Foundations of Human Sociality: economic experiments and ethnographic evidence from fifteen small-scale societies.* Oxford: Oxford Univ. Press.

Henrich, N., & Henrich, J. (2007). *Why humans cooperate: A cultural and evolutionary explanation.* Oxford: Oxford Univ. Press.

Hirschman, A. O. (1970). *Exit, voice, and loyalty: Responses to decline in firms, organizations, and states.* Cambridge: Harvard Univ. Press.

Luckmann, T. (1970). On the boundaries of the social world. In M. A. Natanson (Ed.), *Phenomenology and social reality. Essays in memory of Alfred Schutz* (pp. 73–100). The Hague: Nijhoff.

Mamadouh, V. (1999). Grid-group cultural theory: an introduction. *GeoJournal,* 47(3), 395–409.

Merl, S. (1985). *Die Anfänge der Kollektivierung in der Sowjetunion: Der Übergang zur staatlichen Reglementierung der Produktions- und Marktbeziehungen im Dorf, 1928–1930.* Veröffentlichungen des Osteuropa-Institutes München. Reihe Geschichte: Vol. 52. Wiesbaden: O. Harrassowitz.

(1990). *Bauern unter Stalin: Die Formierung des sowjetischen Kolchossystems, 1930–1941.* Osteuropastudien der Hochschulen des Landes Hessen. Reihe I, Giessener Abhandlungen zur Agrar- und Wirtschaftsforschung des europäischen Ostens: Vol. 175. Berlin: Duncker & Humblot.

(1999). Einstellung zum Privateigentum in Rußland und der Sowjetunion. In H. Siegrist & D. Sugarman (Eds.), *Kritische Studien zur Geschichtswissenschaft: Vol. 130. Eigentum im internationalen Vergleich. 18.-20. Jahrhundert* (pp. 135–160). Göttingen: Vandenhoeck & Ruprecht.

Montgomery, D. R. (2007). *Dirt: The erosion of civilizations.* Berkeley: Univ. of California Press.

Schröder, M. (2009). Integrating welfare and production typologies: How refinements of the varieties of capitalism approach call for a combination of welfare typologies. *Journal of Social Policy,* 38(01), 19–43.

Schulze, E. (2002). Warum blieb in Russland die duale Struktur von Großbetrieb und Hauswirtschaften erhalten? *German Journal of Agricultural Economics (Formerly: Agrarwirtschaft),* (6), 305–317.

Schwartz, S. H. (1970). Moral decision making and behavior. In L. Berkowitz & J. Macaulay (Eds.), *Altruism and helping behavior*, (pp. 127–141). London: Academic Press.

(1994). Are there universal aspects in the structure and contents of human values? *Journal of Social Issues*, 50(4), 19–45.

Schwarz, M., & Thompson, M. (1990). *Divided we stand: Redefining politics, technology, and social choice*. Philadelphia: Univ. of Pennsylvania Press.

Stern, P. C. (2000). Toward a coherent theory of environmentally significant behavior. *Journal of Social Issues*, 56(3), 407–424.

Stern, P. C., Dietz, T., Kalof, L., & Guagnano, G. A. (1995). Values, beliefs, and proenvironmental action: Attitude formation toward emergent attitude objects. *Journal of Applied Social Psychology*, 25(18), 1611–1636.

Thomas, W. I. (1928). *The child in America; Behavior problems and programs*. New York: Johnson.

van den Bergh, J. J., & Stagl, S. (2003). Coevolution of economic behaviour and institutions: towards a theory of institutional change. *Journal of Evolutionary Economics*, 13(3), 289–317.

Veblen, T. (1990). *The place of science in modern civilization and other essays*. New Brunswick: Transaction Publishers [1919].

Verweij, M. et al. (2006). Clumsy solutions for a complex world: the case of climate change. *Public Administration*, 84(4), 817–843.

Wilkinson, R., & Pickett, K. (2009). *The spirit level: Why more equal societies almost always do better*. London: Allen Lane.

Wittfogel, K. A. (1957). *Oriental despotism; a comparative study of total power*. New Haven: Yale Univ. Press.

Worster, D. (1979). *Dust Bowl*. Oxford: Oxford Univ. Press.

4 In a nutshell: the concept of eco-cultural coevolution

Contrary to classic approaches from environmental sociology which are caught between the two poles of unidirectional realism or constructivism, the concept of eco-cultural coevolution combines theoretical considerations from environmental sociology and ecological economics and – in doing so – takes a mediating position. Physical nature and society are not treated as two opposed entities, but are conceptualized as an indivisible unit referred to as eco-cultural habitat. Eco-cultural habitats are specific constellations between human beings and their physical surroundings which are mediated via technologies, institutions, and cultural preferences. These relationships are neither ecologically nor culturally determined but emerge in a dynamic coevolutionary process during which suitable fits between nature and technologies, technologies and institutions, and institutions and culture emerge in a twofold way. This means that these fits have to fulfill both functional necessities (for example, in terms of engineering science or economic functionalism) and ideational preferences.

Describing and explaining the emergence of eco-cultural habitats from an ideal-typical perspective, cultural preferences are initially determined by ecological circumstances rewarding or penalizing reputation building and cooperative behavior. Thereby, theoretical considerations from game theory and transaction cost analysis in combination with arguments from economic functionalism (discussing the advantages of labor division, economies of scale, and allometry) can be employed to elaborate what combinations of physical nature, technologies, institutions, and cultural preferences are the most appropriate. As a general rule, it is argued that cultural preferences favoring cooperative action emerge when the benefits of cooperation outweigh its costs. In contrast, cultural preferences for solitary action prevail when the costs of cooperation outweigh its advantages.

As it is cognitively and societally more efficient to apply one and the same set of cultural preferences to as many realms of social life as possible (as compared with situational decision making or juggling with many object-specific heuristics), cultural preferences tend to spread out and to permeate more and more realms of society as time goes by. This also means that physical nature and society are constituted along similar lines within one and the same eco-cultural habitat. Over the course of time, and via social learning and cultural inheritance, cultural

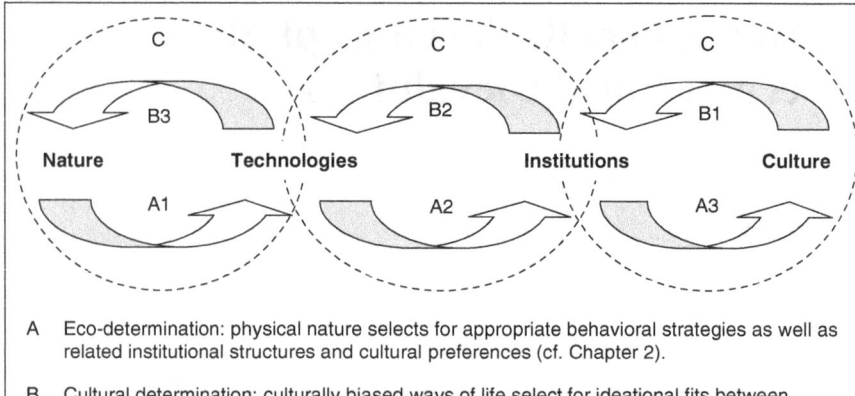

A   Eco-determination: physical nature selects for appropriate behavioral strategies as well as related institutional structures and cultural preferences (cf. Chapter 2).

B   Cultural determination: culturally biased ways of life select for ideational fits between themselves on the one side and suitable institutions, technologies, and physical environs on the other (cf. Chapter 3).

C   Eco-cultural coevolution: twofoldly best adapted matches between nature, technologies, institutions, and culture simultaneously emerge.

Figure 4.1 Ideal-typical illustration showing the emergence and perpetuation of eco-cultural paths of adaptation
Source: compiled by the author.

preferences thus become more and more detached from their initial (functionally determined) origin. They develop a life of their own and become increasingly institutionalized and persistent. At an advanced level of habitat formation, this results in self-reinforcing cultural path-dependencies: socio-ecological tasks and challenges are selected and redefined in such ways as to comply with path-internal views of society and physical nature and – for that reason – become process-able by habitat-specific technologies and institutions. Thus, it could be argued that the initial dynamics of trial and error, during which suitable fits between physical nature and society emerge, gradually make room for rather rigid cultural path-dependencies.

In consequence, and as illustrated in Figure 4.1, adaptational capacities of eco-cultural habitats are affected by the complementary constituents of eco-determination on the one side and – once hegemonic cultural guiding prin-ciples have become established – of cultural persistency and selectivity on the other. This also means that functional necessities may be culturally overridden. However, as long as the culturally preferred way of life can be sustained eco-logically (for example, by technological innovations), mismatches are of minor importance. Only when culturally biased ways of life become ecologically unsus-tainable is continued habitat existence critically endangered, for example, when ecological limits or functional necessities are totally ignored, blocked out, or refused acceptance on the grounds of cultural or ideological reasons. Such con-stellations manifest themselves in individual (or societal) distress, for example, in famine, illness, or death, and are hence supposed to exert increasing pressure to adapt (for example, with regard to novel ideational guiding principles, technologies, or just by selecting the exit option and migrating from one eco-cultural habitat to another).

Against this background and from an ideal-typical perspective, all varieties of eco-cultural paths of adaptation can be located on a continuum which spreads between the two extreme poles of either unregulated, laissez faire-like conditions on the one side and hierarchic despotism on the other. Liberal and coordinated welfare and production regimes lie somewhere in between, with liberal regime types gravitating toward the pole of laissez faire-like conditions, whereas coordinated regimes are more likely to be drawn in the opposite direction. Each path is characterized by distinct world views defining a normatively preferred way of life and – in doing so – stipulating normative rules and ideological superstructures regulating and legitimizing social relationships as well as human interference with physical nature. Liberal habitats are populated by social actors favoring individualistic value orientations, solitary action, and socially unconnected technologies which can be individually and spontaneously applied. As a result of rather low time preferences, inhabitants of liberal habitats do not care much about the welfare of fellow men or environmental sustainability but instrumentalize both for the sake of individual short-term utility. Accordingly, the liberal institutional environment is predestined to coordinate social action in rather laissez faire-like conditions and by competitive means. In contrast, coordinated habitats are populated by social actors favoring collectivistic value orientations, cooperative behavior, and socially connected technologies which can preferably be applied by centralized regulation and long-term planning. Thus, strong time preferences prevail, at least when highly valued social welfare and ecological sustainability are concerned. Consequently, this institutional environment is predestined to facilitate social action in constellations which demand long-term planning and centralized coordination. These forms of path-specific specialization are bought at the price of distinct vulnerabilities, with liberal habitats especially vulnerable to socio-ecological challenges that go beyond individual capabilities and demand for long-term planning, and hierarchic coordination and coordinated habitats particularly vulnerable to socio-ecological challenges which demand an immediate response and short-term flexibility.

Taken together, the combination of considerations from environmental sociology and ecological economics under a coevolutionary theoretical umbrella promises to allow for more fine-meshed and more detailed descriptions and explanations of socio-ecological phenomena (for example, different land and welfare regimes or various approaches to social and ecological crises). Above all, this is facilitated by conceptualizing the process of habitat construction as a dynamic process during which appropriate fits between nature, technologies, institutions, and culture emerge in such a way that both functional necessities and cultural preferences are accounted for. Therefore, the concept of eco-cultural coevolution also promises to allow for highly specific predictions regarding the compatibility between, for example, new technologies and existing cultural preferences and institutional structures (at least as long as the eco-cultural path in question is stable). Further, and due to the fact that culturally biased ways of life permeate all realms of path-specific socio-ecological reality, analytical and predictive capacities should not be confined to either micro or macro analyses, but should be applicable to different spatial and social scales, be it nation states, cities, communities, or individual households.

The extent to which the concept of eco-cultural adaptation can come up to these promises is tested in the following case studies.

5 Rural and urban paths of environmental adaptation and their metabolic characteristics

Against the theoretical background of eco-cultural adaptation as developed in Chapters 2 and 3, the present chapter claims that rural and urban settlement structures and related modes of life represent distinct varieties of eco-cultural adaptation. Therefore, rural and urban settlement structures should be characterized by different adjustments between human beings and their physical surroundings. Further, it is alleged that these differences can be captured empirically by quantitative cross-sectional analyses. The present chapter aims to show that some matches among technologies, institutions, and culturally biased ways of life – more frequently than by random chance – appear in rural rather than in urban settlement structures (and vice versa). Thus, a snapshot identifying contemporary constellations among specific physical environments, technologies, institutions, and cultural preferences in different settlement structures is carved out from the empirical data.

This endeavor raises key issues concerning the operationalization of eco-cultural paths of adaptation: which parts and dynamics of the eco-cultural fabric as presented in Figure 4.1 can be measured and described by what cross-sectional data are available? Here, the concept of social metabolism as described by Marina Fischer-Kowalski (1997) and Sieferle et al. (2006), combined with the idea of different modes of life as presented by Hartmut Häußermann and Walter Siebel (1996), is of special interest. Rural and urban paths of adaptation as distinct modes of life are accompanied by specific social metabolic regimes, which, in turn, become observable in specific adjustments between physical nature, technologies, institutions, and cultural preferences. As will shortly become apparent, technological and cultural properties of eco-cultural paths of adaptation particularly lend themselves to an operationalization in quantitative terms.

In regard to the term "social metabolism", this is defined as "the interaction between human societies and their physical environment" (Sieferle 2011: 323) in the broadest sense. In order to survive, and analogously to cells or animals, human societies have to maintain material and energetic exchange relations with their physical environment. Raw materials have to be extracted and processed into nourishment or other products which are transformed into, for example, physical infrastructure such as houses and streets or other durable goods like cars or furniture. After some time, these goods turn into waste and emissions which are deposited in physical nature (Sieferle et al. 2006: 3). Thereby, metabolic regimes especially differ from one another with regard to their primary source of energy, for example, in terms of land use regimes (extraction of raw materials and energetic inputs) and related forms of organizing social life (for example, in terms of what technologies are available, the occurrence of cooperative advantages (labor division), and related behavioral strategies). In the context of this chapter, the following two metabolic regimes are of special interest: the agrarian metabolic regime based on solar energy stored in biomass (for example, in timber or potatoes) and the urban-industrial metabolic regime based on fossil fuels such as coal, gas, and oil.

According to Sieferle et al. (2006), and due to the fact that the collection and storage of solar energy could hardly be increased in the ancient agrarian metabolic regime, socio-economic development was subject to tight limits. Urban living was a rare exception because agricultural productivity usually did not suffice to support unproductive agrarian and therefore parasitic modes of life such as urbanism (Sieferle et al. 2006: 41). Tapping the "subterranean forest" (i.e. solar energy that had been accumulated and stored in biomass over millions of years) finally sounded the bell for the newly emerging urban-industrial metabolic regime. As a consequence, human life became more and more emancipated from the collection of solar energy via surface area. That, in turn, resulted in drastic social change, especially in industrialization and migration from the country to the cities (Sieferle 2010). Thus, the agrarian metabolic regime was successively replaced by the urban-industrial one. However, and taking seriously the idea of eco-cultural path-dependency, it should be possible to show that some of the following key characteristics of the agrarian metabolic regime should have survived and can still be identified.

Recollecting the theoretical discussion in Chapters 2 and 3, and having in mind these fundamental differences between the agrarian and the urban-industrial social metabolism, it is argued that each metabolic regime is characterized by distinct matches between nature, technologies, institutions, and cultural preferences which can be summarized in the following way: as elaborated by Sieferle et al. (2006), the ancient agrarian metabolic regime is primarily characterized by subsistence farming. Farming activities, again, are mostly characterized by low cooperative advantages, meaning that solitary action is the superior behavioral strategy (cf. subchapter 2.1, Ellickson 1993). This also means that agrarian metabolic regimes are usually accompanied by socially unconnected technologies which can be spontaneously applied by individual actors or small collectives. Cooperation and solidarity are restricted to kin or group. Thus, there is little need to invest in good reputation beyond one's own clan or to cooperate with strangers which, in turn, gives rise to rather individualistic overall cultural preferences (Henrich et al. 2004). In addition, and due to low labor division, subsistence farming is characterized by small-scale production and local trade. Cooperation and exchange relations are based on highly personalized and frequent interaction as well as on cultural homogeneity. Thus, there is no need for third-party enforcement (North 1990). Trying to summarize agrarian metabolic regimes, one could conclude that the most central characteristics are spatial dispersion (following the necessity of collecting solar energy via surface area), physically unconnected technologies (due to dispersion and low cooperative advantages), and the accomplishment of social reproduction within family or clan (face-to-face solidarity).

By contrast, urban-industrial social metabolism does not cover its energetic needs from limited agrarian productivity but from what is believed to be the infinite potential of fossil fuels. On this basis, comparatively high degrees of specialization and labor division as well as urban dwelling become possible. Instead of self-sufficiency, city-dwellers largely depend on cooperation with strangers (expanding time and space). Social reproduction no longer depends primarily on cooperation amongst kin or clan, but can also be accomplished via paid labor and market exchange. As it becomes more or less impossible to sustain frequent and personalized exchange relations in all realms of urban life, it is usually accompanied by some kind of third-party enforcement facilitating exchange relations in the absence of trusted relationships (North 1990). For that reason, urban settlement structures should allow for higher degrees of cultural heterogeneity and less conservatism than do rural settlement structures (where cultural homogeneity is a key precondition for mutual trust and personalized exchange relations). In sum, it could be argued that the most important characteristics of an urban-industrial mode of life are spatial compactness, physically connected technologies, and the accomplishment of social reproduction by means of labor division and market exchange (anonymized solidarity, for example, via social security contributions).

As presented above, rural and urban paths of adaptation and related metabolic regimes represent theoretically (and historically) disjointed cases. Empirically, however, rural and urban modes of life mostly present themselves as mixed types: elements of urban modes of life can also be found in rural areas and vice versa (at least in countries like Germany). Thus, operationalization becomes an intricate endeavor. As indicated above, however, the concept of social metabolism, combined with the idea of different modes of life as described by Häußermann and Siebel (1996), can be used to describe and explain the intricate interaction between energetic, technological, institutional, and cultural peculiarities within rural and urban paths of adaptation. As will be shown, indicators such as different building types (for example, single-family homes or apartment blocks) or the availability of network technologies (for example, gas supply, district heating, and public transport vs. oil heating and high dependence on motorized private transport) are well suited to capturing the technological and energetic characteristics of rural and urban settlement structures (cf. subchapter 5.1). Cultural peculiarities (for example, in terms of traditionalism, familialism, or political conservatism) can be measured by variables such as marital status, number of children, personal education, or political orientation (cf. subchapter 5.2).

However, technological properties and cultural characteristics do not just coexist side by side in isolation, but interact with one another. That, in turn, results in unintended byproducts of rural and urban modes of life, especially with regard to energy consumption and carbon dioxide (CO_2) emissions. Thus, reconstructing specific patterns of energy consumption and the emergence of related CO_2 emissions can provide much information about how technological artifacts such as specific heating systems or building types (and related mentalities) spatially correspond with one another. In addition, distinct patterns of energy consumption and related CO_2 emissions provide key insights into the ecological impacts associated with rural and urban ways of life.

Against the background of these considerations, the following subchapters (5.1 to 5.3) first and foremost concentrate on the correspondences between dwelling technologies on the one side and related modes of social reproduction and respective cultural preferences on the other. The leading assumption is that typical and recurring patterns between dwelling technologies, different modes of social reproduction, and cultural properties (and related patterns regarding CO_2 emissions) can be found on different spatial scales (for example, municipalities or cities).

But why should this be the case? The approach of eco-cultural adaptation claims that culturally biased ways of life and related matches between physical nature, technologies, institutions, and cultural preferences permeate all realms of society. Against this background, and inspired by Mandelbrot's (1967, 1987) studies about self-similar geometric patterns on different analytical scales, it is assumed that self-similar matches between physical nature and society should also be observable on different spatial scales. This should be the case because the universally valid laws as described by allometry, economies of scales, or labor division are always at work – no matter whether rural or urban settlement structures occupy center stage (even though different levels of intensity are observable in rural and urban areas).

By making use of nationwide (individual) data from the renowned German Socio-Economic Panel (GSOEP), spatial patterns in dwelling technologies and related CO_2 emissions are identified in a first step in subchapter 5.1.[1] Second, and based on these analyses, census-like (aggregated) data from all 2056 Bavarian municipalities are used to analyze whether the spatial patterns in dwelling technologies identified in subchapter 5.1 actually correspond with cultural preferences and modes of social reproduction in theoretically expected ways (subchapter 5.2). Last but not least, and using self-collected (individual) data from Munich (Bavaria) and Bolzano (Italy), subchapter 5.3 aims to show that the technological and cultural differences characterizing rural and urban modes of life and related metabolic properties can also be found within urban settlement structures (as suggested by the concept of fractionalization). In subchapter 5.4, the central findings of this research will be collated and critically discussed.

5.1 Spatial correspondences between settlement patterns, dwelling technologies, mobility, and CO_2 emissions at the national level (Germany)

Over recent years and triggered by the increasing global trends of urbanization and climate change, more and more scientists from various backgrounds have contributed to an increasing stock of knowledge about the socio-economic properties of rural and urban modes of life and related environmental impacts (in terms of qualitative and quantitative differences in rural and urban social metabolism). Most studies dealing with this topic concentrate on North America and compare CO_2 emissions of cities on a global scale (for example, Bettencourt et al. 2007, VandeWeghe/Kennedy 2007, Glaeser/Kahn 2010, Glaeser 2011, Hoornweg et al. 2011). Comparing rural and urban residential areas, these studies conclude that urban infrastructural settings enable more efficient ways of life (in terms of energy use) than do suburban and rural structures. Thereby, it is implicitly assumed that energy consumption and carbon emissions result from specific infrastructural settings which are characterized by different forms of physical concentration (for example, in the form of apartment buildings or public transport). In doing so Bettencourt et al. (2007) and Jared VandeWeghe and Christopher Kennedy (2007), among others, implicitly conceptualize ecofriendly (or unfriendly) behavior and resource consumption as unintended side effects of mostly non-reflective day-to-day routines and practices that take place in given (infra-)structural settings which are normally beyond the influence of single individuals (meaning that daily routines and practices are more or less unconsciously guided by these structural properties).

With this, the present chapter aims to determine, using a quantitative approach, whether the general findings mentioned above – namely that urban eco-cultural paths of adaptation are energetically more efficient than rural ones – also hold true for Germany. Furthermore, and in contrast to North American colleagues referred to above, the present analysis concentrates not only on the positive effects of size in terms of physical concentration (for example, in the form of apartment buildings (cf. allometry)), but also emphasizes the importance of social concentration via household size (cf. economies of scale) and its positive environmental effects (Cole/Neumayer 2004). In doing so, the focus is on countervailing effects of social vs. physical concentration that have not yet been taken into account. Regarding urban CO_2 emissions, the environmentally positive effects of physical densification could be counteracted by the high number of single households within urban settlement structures.

In addition, and totally in line with the basic idea of physical concentration, the availability of fuel-transporting network technologies (for example, gas or district heating) in rural and urban settlement structures is also taken into account. That is an important but widely neglected aspect in the discussion about the properties of physically connected technologies and their influence on CO_2 emissions: different fuel types (for example, coal vs. gas) go hand in hand with specific carbon contents and are spatially distributed in systematically different ways.[2] As will be empirically shown for Germany, the distribution of fuel types between urban and rural areas systematically differs (because of the availability of network technologies in cities) and therefore significantly moderates the amount of per capita CO_2 emissions in different infrastructural settings. Against the background of these considerations and in contrast to the studies referred to above, it is assumed, and tested for, that there are no significant differences in energy consumption and related CO_2 emissions between rural and urban settlement structures in absolute terms (due to countervailing effects of physical and social concentration within rural and urban areas), but also that there is considerable qualitative variation in the way these CO_2 emissions originate.

5.1.1 Allometry, scaling, and social metabolism

David Owen (2009), in his book *Green metropolis: why living smaller, living closer, and driving less are keys to sustainability*, describes Manhattan as an ecological utopia. As early as in the second paragraph, he states:

> Most Americans, including most New Yorkers, think of New York City as an ecological nightmare, a wasteland of concrete and garbage and diesel fumes and traffic jams, but in comparison with the rest of America it's a model of environmental responsibility. In fact, by the most significant measure, New York is the greenest community in the United States.
>
> Owen 2009: 1 ff.[3]

Intuitively understandable, Owen reasons that the ecofriendly performance of New York can be explained by its high physical density and compactness.

Recollecting the theoretical arguments of Chapter 2, advantages of physical compactness can be explained by well-established allometric laws and the occurrence of economies of scale (cf. Figures 2.1 and 2.2). In building physics, the advantages of compactness are reflected in the surface area-to-volume (SA:V) ratio, which is defined as the amount of surface area per unit volume of an object. It presents an important indicator in calculating the energy demand for heating. On squaring the surface area of a given object, its volume increases cubically. In other words, the SA:V ratio is inversely proportional to size for a given shape (Galilei 1730). Consequently, and compared with single-family houses, the surface area of apartment buildings is comparatively smaller and less energy is wasted (per unit area). Thus, dwelling in physically compact buildings can be described as a physically connected cultural technique which yields cooperative advantages from higher resource efficiency. Consequently, it can be expected that carbon emissions per capita will reduce with increasing building density. Because apartment buildings are more common in urban than in rural areas (as can be seen in Table 5.1), the urban mode of life should be energetically more efficient (at least with regard to housing).

The general rules from evolutionary biology and statistics regarding the relationships among body mass, body size, anatomy, and metabolic rate can be partly transferred to urban modes of life and the urban metabolism (Bettencourt et al. 2007).[4] In regard to mammals generally, doubling the body mass increases the metabolic rate only by approximately $\beta = 0.75$, and the same holds true for the cardiovascular system. Consequently, the more body mass gained by a mammal, the slower its metabolic rate and pace of life. One of the best documented examples is the idea that an elephant is just a scaled-up ape which, in turn, is described as

Table 5.1 Spatial distribution of key variables determining CO_2 emissions in rural and urban residential areas in Germany

	Size of municipal district (number of inhabitants)				
	<5,000	*5,000–20,000*	*20,000–100,000*	*100,000–500,000*	*>500,000*
Singles (relative frequency)	0.09	0.10	0.12	0.15	0.21
Persons per household	3.01	3.04	2.87	2.72	2.52
Cars per capita	0.55	0.56	0.53	0.47	0.44
Mileage per capita and year/household (1,000 km)	7.55	8.01	7.21	6.49	5.83
Building type					
Agriculturally used residential buildings	0.084	0.039	0.02	0.01	0.003
Detached single- and two-family houses	0.54	0.50	0.31	0.16	0.10
Single- and two-family row houses	0.15	0.19	0.21	0.16	0.14
Apartment buildings (3–4 flats)	0.08	0.10	0.12	0.13	0.10
Apartment buildings (5–8 flats)	0.07	0.12	0.18	0.30	0.31
Apartment buildings (≥9 flats)	0.06	0.05	0.15	0.24	0.35
Heating source					
Oil	0.44	0.50	0.30	0.18	0.23
Gas	0.33	0.37	0.51	0.51	0.44
District	0.06	0.02	0.11	0.23	0.24
Electricity	0.09	0.08	0.06	0.06	0.08
Coal/timber	0.20	0.14	0.07	0.04	0.03
Living space per capita (m²)	39.99	43.81	40.44	38.51	40.76
Household income per capita (1,000 €/month)	1.12	1.34	1.32	1.37	1.52
CO_2 emissions per capita (t/year)					
Mobility	1.44	1.57	1.43	1.29	1.20
Housing	3.36	3.42	3.02	2.85	3.10
Mobility and housing	4.82	4.95	4.40	4.11	4.24

Source: compiled by the author. GSOEP 1998 and 2003.

a scaled-up mouse (Haldane/Maynard Smith 1985). In accordance with the SA:V and mass-to-metabolism ratios, Bettencourt et al. (2007) were able to show that similar economies of scale can also be observed in cities, but only with regard to their material infrastructure. They report that by doubling the population of a city, its material infrastructure (for example, the length of electrical cables or wastewater pipes) only increases by $\beta = 0.8$.

However, and in contrast to mammals, Bettencourt et al. (2007) also report that the urban metabolism does not slow down with increasing size. Instead, city life even accelerates ($\beta = 1.2$). Doubling the population of a city increases the rate of various interactional outcomes (such as the rate of crimes, infections, or innovations) by a factor of 1.2 which results in, for example, higher incomes. Theoretically, this acceleration can be explained by cooperative advantages derived from labor division and specialization (cf. Chapter 2) and by the effects of social contagion in the broadest sense. Regarding the overall ecological effects of urban modes of life, it can be assumed that comparably higher incomes and smaller household sizes in cities counteract the positive effects of physical concentration.

As indicated, however, both the SA:V and mass-to-metabolism ratios neglect the importance of social concentration via household size (economies of scale) and its reverse effects with regard to carbon emissions within physically different residential environments. Consequently, and partly in contrast to, for example, Bettencourt et al. (2007) or VandeWeghe and Kennedy (2007), the following contrary effects can be expected: due to higher rents and land prices, urban residents tend to live in compact buildings (for example, apartment buildings) with comparably less living space and surface area per capita.[5] These man-made physical environments result in comparably low energy demands for heating. What is more, city-dwellers do not necessarily depend on cars for transportation, but can make use of public transport, go by bike, or just walk. Although these facts seem to suggest that cities are energetically more efficient than rural areas, it has to be kept in mind that these aspects of physical concentration might be counteracted by higher incomes (and related consumption patterns) and smaller household size (cf. subchapter 5.2).

The opposite can be observed in rural areas. As a result of their dispersed structure and lower population density[6], physically connected forms of transportation (public transport) do not pay off. Consequently, rural residents are heavily dependent on private transport and cars. In addition, and because of lower rents and land prices, they (still) tend to live in detached single-family homes (higher surface area and living space per capita). However, social concentration via household size and lower incomes could counteract the negative impacts of physical dispersion.

In addition, and due to different degrees of physical concentration, rural and urban paths of ecological adaptation systematically differ in their use of energy carriers. Whereas heating with oil is often found in rural areas, the more carbon-saving district- and gas-heating technologies are widespread in urban regions. Urban structures comply with the requirements of physically connected network technologies (for example, connected distribution systems for gas or district heating), whereas dispersed rural settlement structures call for uncon-nected energy supplies (for example, via privately owned oil tanks). In short, there are strong theoretical arguments that urban modes of life and related infra-structures facilitate more energy-efficient ways of housing and transportation than rural structures.

5.1.2 Settlement patterns and energy consumption in the GSOEP

The present analyses are based on the German Socio-Economic Panel (GSOEP). The panel was established in 1984 and samples almost 11,000 households and more than 20,000 individuals annually (DIW 2011), and consequently it provides a broad range of socio-economic data. Within the scope of this chapter, the sur-vey waves of 1998 and 2003 are of special interest. In contrast to other waves, these provide crucial information on the spatial distribution of energy technolo-gies, energy consumption, and related carbon emissions at the household level. On this basis, it is possible to measure the properties of rural and urban paths of adaptation and related metabolic patterns of energy consumption as a function of different technologies, building physics, city size, and various socio-economic, infrastructural, and geographic factors of influence while controlling for differ-ences at the household level.

With regard to housing, the respondents were questioned in depth about their expenses for heating and electricity. In addition, detailed information is avail-able on individual and household characteristics (for example, household size, building type, living space, size of the residents' village or town, and household income). With regard to mobility, respondents were asked about the number of cars available, fuel type, annual mileage, and gasoline consumption per 100 km.[7]

After identifying all relevant variables (both theoretically and practically within the GSOEP sample), the dependent variable (total annual carbon emissions per capita related to heating and mobility) was constructed. All energy expenses (for example, for gas, oil, coal, district heating and mileage) were translated into abso-lute energy amounts and carbon emissions. With regard to housing and using the respective energy prices for each energy carrier and year (Federal Ministry of Economics and Technology 2011), energy expenses were converted to absolute amounts of energy which were then translated into carbon emissions. With regard to private transport, total carbon emissions were calculated from annual mileage, gasoline consumption per 100 km, and fuel type. In both cases, different carbon contents of each energy carrier were taken into account using the total direct and indirect emission factors (Fritsche 2007, Fritsche/Rausch 2007, Gemis 2011).[8] The total carbon emissions from housing and mobility finally gave the dependent variable.

The dependent variable (CO_2 emissions) is of interest here not only in absolute terms. Although total amounts of CO_2 emissions are crucial – for example, in order to evaluate the overall ecological impacts of rural and urban paths of adap-tation in the realms of housing and mobility – qualitative differences in the way these emissions appear in rural and urban paths of adaptation are also relevant in

Table 5.2 Carbon emissions: how much does social and physical concentration actually explain?

2003	Mobility			Housing			Mobility and housing		
	1	2	3	1	2	3	1	2	3
Population of municipal district (reference: <5,000)									
5,000–20,000	127.11*** (33.10)	-10.02 (31.38)	11.32 (28.48)	-33.17 (77.22)	-101.88 (75.44)	-107.01 (66.14)	-75.63 (100.07)	-152.16 (99.04)	-119.29 (87.95)
20,000–100,000	-22.17 (30.19)	-28.96 (29.23)	-28.09 (26.65)	-394.95*** (74.96)	-182.73* (76.58)	-80.33 (67.36)	-527.37*** (93.39)	-310.07** (94.80)	-157.74* (83.07)
100,000–500,000	-139.84*** (35.25)	-70.84* (34.50)	-82.05* (32.31)	-749.15*** (67.05)	-368.81*** (66.73)	-184.44*** (55.65)	-932.57*** (91.82)	-553.64*** (90.61)	-266.16*** (77.13)
≥500,000	-214.95*** (37.33)	-133.60** (42.61)	-153.23*** (36.34)	-708.33*** (76.86)	-370.62*** (76.15)	-100.22 (65.23)	-1014.41*** (102.06)	-703.32*** (100.27)	-252.16** (87.03)
Household income per capita		0.50*** (0.06)	0.37*** (0.05)	1.13*** (0.05)	1.08*** (0.05)	0.27*** (0.04)	2.09*** (0.08)	2.05*** (0.08)	0.91*** (0.08)
German nationality (1 = not German)		84.71** (27.13)	-104.31*** (25.33)	353.82*** (61.69)	410.63*** (58.33)	-71.02 (51.08)	388.25*** (77.36)	473.29*** (74.34)	-243.92*** (66.37)
Age		-5.19*** (0.92)	-12.19*** (0.71)	25.46*** (1.74)	27.01*** (1.66)	8.06*** (1.71)	15.96*** (2.13)	16.35*** (2.05)	-5.11* (2.04)
Cars per household		621.50*** (18.50)	826.55*** (22.94)				489.50*** (39.42)	388.15*** (40.74)	834.07*** (44.29)
Education (reference: secondary I "Hauptschule")									
Secondary II ("Realschule")		133.59*** (27.42)	97.40*** (24.59)	-62.40 (57.24)	-12.90 (55.20)	-31.36 (50.21)	43.49 (74.35)	118.18 (73.30)	23.10 (66.68)

	(1)	(2)	(3)	(4)	(5)	(6)	(7)	(8)
High school ("Abitur")	197.37*** (42.62)	175.69*** (33.64)	-82.33 (62.55)	-64.50 (59.76)	-48.62 (52.47)	-51.07 (84.96)	-10.56 (82.95)	-34.28 (73.77)
Household size (ln)		-951.4*** (42.60)			-989.31*** (70.82)			-1740.19*** (97.04)
Living space per capita					36.11*** (2.35)			39.34*** (3.00)
House: year of construction (reference: houses built before 1918)								
1918–1948			-156.80+ (92.33)	-143.91 (93.35)	-59.65 (86.09)	-132.70 (111.71)	-127.94 (112.82)	-67.21 (103.04)
1949–1971			-258.67*** (72.77)	-293.73*** (71.24)	-114.03+ (64.05)	-222.16* (88.74)	-264.96** (87.25)	-86.92 (78.55)
1972–1980			-368.74*** (82.34)	-431.46*** (81.73)	-199.96** (72.46)	-142.66 (108.56)	-202.05+ (108.61)	17.06 (97.82)
1981–1990			-378.96*** (92.95)	-324.74*** (88.96)	-58.11 (80.05)	-209.20+ (118.76)	-161.65 (115.65)	130.58 (102.93)
1991–2000			-877.96*** (73.80)	-705.56*** (72.37)	-464.71*** (63.13)	-777.74*** (96.20)	-598.52*** (95.58)	-324.65*** (83.95)
2001 or later			-709.26*** (117.48)	-523.06*** (1114.49)	-550.03*** (105.70)	-723.51*** (169.43)	-512.62** (168.78)	-544.37*** (140.94)
Heating (reference: district heating)								
Oil				1421.62*** (84.51)	1065.98*** (79.32)		1516.61*** (99.52)	1025.26*** (92.41)
Gas				204.50** (64.72)	-8.89 (62.57)		235.13** (76.59)	-45.88 (73.11)
Electricity				598.11*** (154.70)	550.56*** (147.10)		600.07*** (167.89)	540.36*** (158.68)
Coal/timber				800.66*** (148.06)	646.26*** (150.84)		800.22*** (170.27)	502.27** (168.59)

Table 5.2 (cont.)

2003	Mobility			Housing			Mobility and housing		
	1	2	3	1	2	3	1	2	3
Building type (reference: detached single- and two-family houses)									
Single- and two-family row houses						-210.20*** (55.94)			-217.62** (80.04)
Apartment buildings (3–4 flats)						-331.55*** (74.37)			-308.83*** (87.68)
Apartment buildings (5–8 flats)						-467.93*** (78.10)			-486.98*** (95.14)
Apartment buildings (≥9 flats)						-656.47*** (72.92)			-742.15*** (94.50)
Constant	1480.96*** (21.95)	148.47** (46.95)	1390.34*** (70.30)	1194.49*** (121.22)	356.19** (123.79)	2227.72*** (210.85)	1366.29*** (159.23)	622.14*** (159.61)	3202.04*** (257.93)
N	21,193	19,347	19,347	8,460	8,460	8,420	8,002	8,002	7,964
R^2	0.005	0.296	0.353	0.186	0.253	0.406	0.280	0.321	0.467
Adjusted R^2	0.004	0.295	0.353	0.185	0.252	0.405	0.278	0.319	0.465

Note: simple linear regressions with carbon emissions (kg CO_2 per capita and year) as dependent variable. Robust standard errors in parentheses. * $p \leq 0.05$, ** $p \leq 0.01$, *** $p \leq 0.001$.

Source: compiled by the author. GSOEP 2003.

order to learn more about eventual differences in rural and urban dwelling technologies and related modes of life.

Corresponding to this central focus of research, simple linear regression models with robust standard errors were run (Greene 2003; Gujarati 2003). To gain deeper insights into the intricate links between settlement structures, technologies, and energy demand and to learn more about the specific influences of the relevant independent variables (for example, household size or living space), separate simple linear regression models for 1998 and 2003 are presented.[9] After some descriptive statistics have been presented, the analysis is then divided into four main parts (cf. Table 5.2): first, explaining carbon emissions caused by private transport (Mobility); second, explaining carbon emissions triggered by building operations (Housing); third, scrutinizing the combined effects of mobility and housing (Mobility and housing); and fourth, checking for the robustness of these findings with fixed effects models for the GSOEP waves of 1985 to 2009 with monthly utilities for housing as the dependent variable in Model 4, cf. Table 5.3 (Allison 2009, Angrist/Pischke 2009).[10] To determine the specific influence of relevant independent variables, each block of the following simple linear regression models is divided into three sub-models which are then gradually enlarged. By enlarging the models step by step, the effects of the newly introduced variables can be observed. The following discussion concentrates on the most interesting models and most significant effects.

5.1.3 Explaining spatial differences in the generation of CO_2 emissions

As one looks for systematic differences in dwelling technologies between rural and urban paths of adaptation, some descriptive statistics seem to offer an appropriate way to gain a general overview of the spatial distribution of dwelling technologies, as well as other central variables influencing energy consumption and CO_2 emissions (cf. Table 5.1). Rural and urban paths of adaptation are distinguished by number of inhabitants living in a given municipal district, ranging from fewer than 5000 in rural areas to cities with more than 500,000. As indicated in the working hypothesis, two distinct forms of concentration within urban and rural areas can be observed. In rural areas, social concentration takes place via household size, whereas – as indicated by the high number of detached single- and two-family houses – there is little physical densification. The converse is evident in urban settlement structures: these are characterized by a high degree of physical concentration via apartment buildings and a high rate of single households.

As theoretically expected, physically dispersed rural settlement structures are accompanied by unconnected structures of energy supply, dominated by oil and coal heating. In physically connected urban residential areas, in contrast, network technologies (gas and district heating) are more common. In short, fuels with a higher carbon content per unit of energy are more commonly used in areas with dispersed settlement structures, where network technologies are barely profitable or cannot be realized at all.[11] As will be seen below, a substantial portion of cities' savings potential with regard to carbon emissions can be explained by the use of district or gas heating rather than oil or coal. As expected, mileage resulting from private transportation declines and household income increases with rising levels of physical connectedness. However, what cannot be maintained (at least in the GSOEP sample) is the assumption of significant differences in per capita living space between rural and urban residential areas resulting from differences in rents and land prices. Although differences do exist, these appear to be rather insignificant at first glance.

In regard to the dependent variable – carbon emissions from housing and private transport – the range of observable path- and settlement-specific averages (min. 4.1 t CO_2/capita; max. 4.9 t CO_2/capita) is in accord with the general figures reported for Germany in the relevant literature, which show that roughly 50% of the carbon emissions of a German average consumer (total 9.7 t CO_2/capita) can be explained by housing and mobility (Weber/Perrels 2000).[12] There are also clear and recognizable differences in per capita carbon emissions between compact and dispersed forms of residential areas. Although the data suggest that there is no clear-cut linear relation between settlement structures and carbon emissions, the analysis presented here shows that condensed infrastructural settings result in energetically more efficient patterns of private transport and housing. Even though the differences between German urban and rural settlement structures are (in absolute and relative terms) not as pronounced as in the case of Toronto, the findings presented here are in line with those of VandeWeghe and Kennedy (2007) who report averages of 6.42 t CO_2/capita for the center of Toronto and 7.74 t CO_2/capita for the surrounding districts. Furthermore, these findings do not justify the conclusion that social and physical concentration totally cancel each other out. Rather, the findings of Edward Glaeser (2011), Bettencourt et al. (2007), and VandeWeghe and Kennedy (2007) seem to be valid for Germany, even though the differences between rural and urban settlement structures seem to be weaker in Germany (at least in the GSOEP sample and providing only energy expenses are considered).

Next, the reciprocal action between different determinants of energy consumption and carbon emissions will be analyzed by means of simple linear regression models (cf. Table 5.2). Based on these analyses, conclusions regarding potential differences between rural and urban paths of adaptation will be drawn.

Mobility

As only district size was controlled for in the first mobility model, it can be described as the minimum model (reference value: districts with fewer than 5000 inhabitants). Against the background of theoretical considerations and with regard to the dependent variable "carbon emissions from private transportation", increasing saving potentials from rural to urban areas would be expected. However, the minimum model only shows ambiguous results ($R^2 = 0.005$). As further variables are added, the model Mobility 2 provides the theoretically expected effect-direction of district size and gains explanatory power ($R^2 = 0.296$). Moving from a residential area with fewer than 5000 inhabitants to one with more than 500,000 results in an annual saving potential of 133 kg CO_2/capita. Here, the number of cars per household seems to have the greatest impact and strongly moderates the effect of district size. Expanding the model by household size increases the explanatory power of the model by another 5% ($R^2 = 0.353$). Now, only municipal districts with at least 100,000 inhabitants significantly contribute to decreasing carbon emissions.

Here, it could be argued that the minimum number of 100,000 inhabitants presents a threshold level from which the positive effects of size (physical concentration) as theoretically described gain momentum. Where a system of public transportation (connected technology) is close at hand, one may prefer to make use of that instead of using one's own car (unconnected technology). Service utilities are available within walking distance. However, and as theoretically assumed, social concentration (household size) seems to be the most important predictor for carbon emissions from private transportation (followed by the number of cars per household).

Housing

Analogous to mobility and with comparable results, a minimum model for housing was run first (not reported here, $R^2 = 0.008$). Increasing saving potentials in carbon emissions between rural and urban residential areas are revealed only by introducing socio-economic variables such as income and physical variables like building age. Then, increasing gains in saving potentials between dispersed and condensed infrastructural settings can be observed ($R^2 = 0.186$). However, and as the model Housing 2 suggests ($R^2 = 0.253$), these saving potentials nearly halve when the type of fuel used for heating is controlled for (reference value: district heating). Last but not least, district size (used as a proxy for rural and urban paths of adaptation) seems to lose nearly all its explanatory power when building type, living space, and household size are controlled for (cf. model Housing 3, $R^2 = 0.406$).

As theoretically expected (cf. SA:V ratio), saving potentials gradually increase with building size. Compared with detached single- or two-family houses, moving into an apartment building with more than nine flats results in a highly significant decrease in yearly carbon emissions of 656 kg CO_2/capita. Another key source of potential savings is provided by buildings no older than 20 years, probably due to stricter energy standards. The highest savings, however, again emanate from household size and heating fuel, especially district heating (against which the other fuels are controlled for by dummy variables). For example, doubling the household size (which means an increase of 100%) decreases carbon emissions by 989 kg CO_2/capita per year (which is also highly significant). Thus if household size increases from two to three persons, carbon emissions per capita decrease by 684 kg CO_2. To put it in a nutshell and with regard to housing, district size per se does not present sufficient explanatory power. Instead, the most

Table 5.3 Monthly utilities: are the findings robust?

	Model 4: Monthly utilities per capita (in euros)			
	1	2	3	4
Size municipal district (reference: <5,000)				
5,000–20,000	−5.35***	−5.54***	−5.56***	−1.95
	(1.22)	(1.35)	(1.38)	(1.19)
20,000–100,000	−7.61***	−6.61***	−6.73***	3.16**
	(1.10)	(1.21)	(1.24)	(1.05)
100,000–500,000	−21.79***	−22.59***	−22.86***	−1.58
	(1.33)	(1.47)	(1.50)	(1.26)
≥500,000	−25.70***	−22.92***	−23.78***	0.80
	(1.45)	(1.64)	(1.71)	(1.46)
Household income per capita		0.003***	0.003***	0.01***
		(0.0004)	(0.0004)	(0.0004)
German nationality (1 = not German)		35.86***	36.74***	36.33***
		(0.90)	(0.90)	(0.88)
Age		−0.01	0.04	0.31***
		(0.03)	(0.03)	(0.04)
Education (reference: secondary I "Hauptschule")				
Secondary II ("Realschule")		−2.88**	−2.77**	−0.12
		(1.00)	(0.99)	(0.91)
High school ("Abitur")		−12.08***	−11.84***	−4.18***
		(1.17)	(1.17)	(1.04)
House: year of construction (reference: houses built before 1918)				
1918–1948			−3.18***	−0.56
			(0.92)	(0.79)

	(1)	(2)	(3)	(4)
1949–1971			-6.67***	-1.16
			(0.97)	(0.83)
1972–1980			-4.91***	0.29
			(1.07)	(0.94)
1981–1990			-4.64***	-1.09
			(1.13)	(0.98)
1991–2000			-10.71***	-9.34***
			(1.21)	(1.05)
2001 or later			-11.13***	-19.48***
			(1.87)	(1.67)
Household size (ln)				48.59***
				(0.84)
Living space per capita				0.56***
				(0.03)
Building type (reference: detached single- and two-family houses)				
Single- and two-family row houses				-2.90
				(1.16)
Apartment buildings (3–4 flats)				-17.43***
				(0.91)
Apartment buildings (5–8 flats)				-19.84***
				(0.89)
Apartment buildings (≥9 flats)				-21.12***
				(0.98)
Constant	108.86***	77.44***	79.67***	-10.44***
	(0.77)	(1.89)	(2.01)	(2.35)
Observations	450,702	355,723	349,354	347,767
Individuals	49,856	46,260	45,717	45,664
R^2	0.004	0.015	0.016	0.075
Adjusted R^2	0.004	0.015	0.016	0.075

Note: fixed-effects regressions with monthly utilities (euros per capita) as dependent variable. Robust standard errors in parentheses. * $p \leq 0.05$, ** $p \leq 0.01$, *** $p \leq 0.001$.
Source: compiled by the author. GSOEP 1985–2009.

important explanatory variables (which go hand in hand with district size and related modes of life) are household size, the available fuel type used for heating, building type, and building age. Regarding the question of how CO_2 emissions emerge in different settlement structures, the highly significant impacts of house-hold size and fuel type are particularly marked.

Mobility and housing

Analyzing the combined effects of mobility and housing, the last three models generally reproduce the above findings. Once again, the minimum model does not reveal any significant explanatory power or a clear-cut distribution of car-bon emissions between rural and urban residential areas (not reported here, $R^2 = 0.010$). However, introducing socio-economic variables and building age (cf. Mobility and housing 1) yields the expected distribution of saving potentials between urban and rural areas (which nearly doubles from 527 kg CO_2/capita in districts with 20,000 to 100,000 inhabitants to 1014 kg CO_2/capita in cities with more than 500,000 inhabitants). Here again, controlling for the spatial distribution of heating fuels results in a sharp decline in the effect-size of district size, but does not change the general direction of the effect (cf. model Mobility and housing 2, $R^2 = 0.321$). Last but not least, and in general accordance with the models dis-cussed earlier, the third model indicates that a district size of 100,000 inhabitants or more is necessary to yield significant savings in carbon emissions. However, the variables household size, fuel type (especially oil), building type (especially apartments with more than nine flats), and the number of cars per household have a much greater influence on carbon emissions than district size alone. With an explained variability of $R^2 = 0.467$ (541 for 1998), it is possible to explain nearly 50% of all carbon emissions resulting from the operation of buildings and private transportation.

Taken together, explained variability gradually improves when new (theoretic-ally justifiable) variables are introduced into the models. For 2003, the explained variability ranges from $R^2 = 0.01$ to 0.47. Thereby, the effect-size of the inde-pendent variable size of municipal district (used as a proxy to identify possible differences between rural and urban paths of adaptation) gradually declines in all models. However, and although all theoretically reasonable variables that could confound the effect of municipal district size and compactness (for example, building type) to a certain extent are controlled for, the size of municipal dis-trict still significantly influences carbon emissions in the last model (at least in districts with more than 100,000 inhabitants, cf. Mobility and housing 3). Thus, further covariates are probably simultaneously related to district size and carbon emissions.

For example, differences in rural and urban modes of life could account for these remaining differences. Rural areas are characterized by social concen-tration, which is also expressed by daily routines and practices that take place within one's own four walls (for example, inviting people for dinner, preparing meals, and "do it yourself work" instead of using services). In urban settings, these activities and services are widely outsourced. Clothes are given to the laundry, lunch is caught from a takeaway, and meeting friends happens in bars or concert houses (Häußermann/Siebel 1996, Otte/Baur 2008). Consequently, it could be argued that city dwellers spend more time outside their private homes and tend to have a higher demand for energy-consuming services (which of course are not included in their individual bills for heating or electricity). Clearly more research has to be done, especially with regard to the interaction between carbon emissions and household income, all the more because the potential impacts of rebound effects have not yet been accounted for. Even if urban residents use less energy for building operations and private transportation, this does not necessarily imply that they are per se environmentally more efficient. As their energy expenses are comparatively low, their household budgets should allow for more consumption of other goods, which might be either more or less carbon intensive, for example, costly haircuts with lower carbon intensity or air travel with higher carbon intensity (Holden and Norland 2005, Lähteenoja et al. 2008).

In order to control for these assumed individual differences in modes of life and to test whether the positive effect of district size persists after averaging out individual fixed effects, a robustness check using fixed effect models to control for time-constant unobserved heterogeneity was run (cf. Table 5.3). The following variations of Model 4 show how changes in the independent variables are associated with changes in monthly utilities.

Although the results from Tables 5.2 and 5.3 are not directly comparable because they are related to different dependent variables, there is no doubt that the main findings remain remarkably stable. District size has a significant effect on monthly utilities per capita providing household size, living space per capita, and building type are not controlled for. Thus, the effect of district size seems to result from social and physical density. Furthermore, the fixed-effects models also corroborate the findings according to which a district size above 100,000 inhabitants can be seen as an important threshold. Finally, other forms of densification again significantly reduce individual expenses. This holds true for household size, living space per capita, and building type.

5.1.4 Summary

Regarding CO_2 emissions caused by rural and urban dwelling technologies and related structural peculiarities in absolute terms, the analysis presented above confirms the findings of, among others, VandeWeghe and Kennedy (2007) and Bettencourt et al. (2007): significant differences between rural and urban areas in per capita carbon emissions also exist in Germany. However, these variations in CO_2 emissions are explained not only by physical concentration, but also by social concentration (economies of scales) and the spatially uneven availability of fuel types used for heating. Thus, the identified differences in per capita carbon emissions are mainly due to differences in the type of heating used, household size, building type, numbers of cars per household, and living space per capita.

But does the spatial distribution of these variables also follow the theoretically expected pathway?

Urban modes of life are characterized by high degrees of physical compactness and low degrees of social concentration (in terms of household size), whereas the opposite is true of rural paths of adaptation. Physical compactness (for example, in terms of apartment buildings) corresponds with network technologies such as public transport or district heating and comparably smaller households. Physical dispersion (for example, in terms of detached houses) primarily goes hand in hand with larger households and physically unconnected technologies such as private transport by car or heating systems based on grid-unconnected energy carriers such as oil. Thus, social and physical concentration actually counteract one another – at least providing that energy expenditures are exclusively dealt with in the realms of mobility and housing. These findings strongly corroborate the theoretical assumptions presented above, according to which adaptational paths fuelled by solar energy (in combination with comparatively low degrees of labor division) are usually marked by dispersed and physically unconnected technologies, whereas adaptational paths relying on fossil fuels and strong labor division are characterized by high degrees of physical concentration and connected technologies (Ellickson 1993, Sieferle et al. 2006). Of course, and in order to avoid misunderstandings, contemporary rural paths of adaptation depend largely on fossil fuels, too (at least in Germany). However, the data presented above suggest that grid-unconnected fuel types simply slipped into the preexisting dispersed structural properties of the ancient agrarian adaptational path as described by Sieferle et al. (2006) and largely affect its current metabolic characteristics, even at present. In urban areas, as a result of the stronger economies of scales derived from physical concentration, physically connected technologies have become widespread.

The results generated above not only yield empirical evidence concerning different dwelling technologies in rural and urban paths of adaptation, but also provide the first tentative hints regarding corresponding cultural preferences and modes of social reproduction. In this context, it was argued that physical environments yielding low cooperative advantages should be accompanied by low degrees of labor division and specialization (and vice versa), meaning that social reproduction has to be accomplished within the household (more family-centered life forms rather than reproduction via market exchange, as is the predominant case in urban areas). Because rural households – compared with urban households – are characterized by lower per capita incomes and are significantly larger, the data presented above could be read in support of this assumption. Further, and on account of the fact that central peculiarities of the (ancient) rural adaptational paths still seem to exist today – both with regard to household size and (un)connected technologies – it could be argued that the theoretical assumption, according to which eco-cultural coevolution results in persistent habitat-specific path-dependencies, is also valid.

Regarding the overall research questions of this chapter, the presented analysis grants the first key insights: the technological dimension of adaptational paths can be quantitatively operationalized and yields the theoretically expected spatial patterns – at least in the case of dwelling technologies in Germany. Whether this also holds true for other central components of eco-cultural fabric – especially for cultural preferences and related modes of social reproduction, and whether these components spatially correspond with rural and urban dwelling technologies in the theoretically expected way, will be further examined in the following chapters.

5.2 Elective affinities between dwelling technologies, cultural preferences, and social metabolisms at the regional level (Bavaria)

In Chapters 2 and 3, it was argued that physically and socially (un)connected dwelling and heating technologies as identified above should correspond with compatible daily routines and practices (for example, in terms of cooperative behavior and labor division) as well as appropriate cultural preferences (for example, individualistic, conservative, or social-democratic value orientations). Whether this theoretical assumption holds true represents the central point of research of the present subchapter. Based on the findings about the spatial distribution of dwelling technologies and related CO_2 emissions as presented in subchapter 5.1, the present subchapter aims to show that rural and urban paths of adaptation are actually marked by distinct and theoretically predictable matches between dwelling technologies on the one side and different modes of social reproduction, cultural preferences, and related variations in resource consumption on the other. Accordingly, and regarding the theoretical concept of eco-cultural adaptation, not only theoretical questions concerning its predictive accuracy, but also crucial practical issues concerning the operationalizability and measurability of further constituents of the eco-cultural fabric (cf. Figure 4.1) are addressed.

5.2.1 Working hypotheses: bundles of settlement structures, mentalities, and modes of social reproduction

Rural and urban paths of adaptation have been described here as the interplay between distinct metabolic regimes and related modes of life. Combining Sieferle et al.'s (2006) description of agrarian and urban-industrial metabolic regimes with the theoretical considerations regarding the occurrence of cooperative advantages and related behavioral strategies (cf. Chapter 2, Ellickson 1993, Henrich et al. 2004), it was argued that rural and urban paths of adaptation can be characterized in the following way: whereas rural paths of adaptation are characterized by spatial dispersion, physically unconnected technologies, and the accomplishment of social reproduction via "do it yourself work" within family or clan, the most important features of urban paths of adaptation are spatial compactness, physically connected technologies, and the accomplishment of social reproduction via paid labor combined with market exchange. Based on these considerations, the following three working hypotheses regarding the correspondences among dwelling technologies, modes of social reproduction, and cultural preferences are now of special interest.

First, and with regard to the correspondence between dwelling technologies and cultural preferences, it can be assumed that physically compact urban settlement structures coincide with non-conservative and social-democratic value orientations. Rural areas, in contrast, rather show a predominance of conservative-individualistic value orientations.

Second, and considering the interplay among dwelling technologies, cultural preferences, and prevalent modes of social reproduction, urban settlement structures should coincide with high degrees of labor division and specialization. Rural settlement structures should rather correspond with moderate degrees of labor division and specialization. Consequently, it can be assumed that social reproduction is mainly accomplished via paid labor and market exchange in urban areas, whereas social reproduction takes place via "do it yourself work" within the household to a comparably greater extent in rural areas.

Third, it is analyzed how the different combinations of dwelling technologies, cultural preferences, and modes of social reproduction – as problematized in the first and second working hypotheses – are reflected in specific metabolic characteristics. In contrast to subchapter 5.1, where only energy expenditures in the realms of housing and mobility were considered, it is assumed here that rural modes of life result in lower CO_2 emissions than urban modes of life. Such a finding is expected once the ecological impacts of income disparities (between rural and urban areas) and related levels in consumption are taken into account (Hertwich 2005, Holden/Norland 2005).

5.2.2 Operationalizing eco-cultural habitats and related social metabolisms

The subsequent analyses are based on a dataset which was compiled during a triennial research project funded by the German Ministry of Education and Research.[13] This dataset lends itself to the analyses of rural and urban paths of adaptation in that it contains key variables regarding technological, economic, cultural, and metabolic characteristics of all 2056 Bavarian municipalities. To give some examples, the dataset covers the following variables (among others): size of municipality (in terms of inhabitants); living space per capita (differentiated according to building type and heating technology); number of commuters and related mileage; number of cars (differentiated according to engine size and fuel type); average household size and household income; unemployment data; level of educational qualification; balance of migration; election results; and the numbers of weddings and divorces. In order to avoid misunderstandings, and in contrast to subchapter 5.1 (and 5.3), the sampling units here are not individual households but municipalities. Thus, each of the 2056 Bavarian municipalities represents a statistical aggregate of its inhabitants and their socio-economic and demographic characteristics as well as of communal infrastructure properties such as building type, rental prices, or migration balance (across municipal borders).

As the dataset described above mostly contains process-generated data collected by public bodies, the data quality should be better than in the case of regular household surveys. Especially with regard to information which is either sensitive (for example, political orientation or household income), hard to remember, or prone to effects of social desirability, the validity of process-generated data is much higher. Moreover, and thanks to the fact that the data have a comprehensive census-like character, systematic non-response does not occur and there is no need for inferential statistical procedures to draw conclusions regarding the population. Finally, and taking municipalities as sampling units, direct access to valuable information such as election outcomes or the number of commuters (and related mileage) is provided. As will be shown shortly, such aggregate data contain key information about the properties of eco-cultural paths of adaptation that cannot be surveyed on the level of individual households.

Coming back to the three working hypotheses formulated above, the following variables lend themselves to a check for validity (cf. Table 5.4). Analogous to sub-chapter 5.1, rural and urban paths of eco-cultural adaptation can be categorically distinguished by five size groups (whereby assignment depends on total population). As a result of small numbers in the case of municipal districts of more than 100,000 inhabitants ($N = 8$), however, the number of inhabitants per 100 m^2 of constructed area is also used here as a metric proxy for (high or low) physical compactness.

Regarding the first working hypothesis, rural and urban "dwelling technologies" are measured by building type, energy carrier used for heating, living space per capita, and household size. It is expected that rural paths of adaptation will show low levels of physical concentration, grid-unconnected heating sources, and comparably bigger households. In contrast, urban paths of adaptation should be marked by physical compactness in combination with grid-connected heating sources and smaller households. "Cultural preferences", in turn, are measured by election outcomes, the number of children, marital status, religious affiliation, and the rate of women's employment. According to the relevant literature, these variables should provide reliable hints at different degrees of conservatism, familialism, and cultural homogeneity (Häußermann/ Siebel 1996, Otte/Baur 2008). For rural areas, it is expected that detached housing in combination with grid-unconnected energy carriers coincides with a traditional family image (women staying at home to raise children and men as the main breadwinners), high ownership rates, and conservative-possessive individualistic value orientations (Mcpherson 1962). Physically compact dwelling technologies, in turn, are expected to coincide with a higher proportion of tenants and social-democratic value orientations in combination with a broader variety of life and family plans.

Concerning different modes of "social reproduction" as problematized in the second working hypothesis, measuring different degrees of labor division appears to present a promising approach. Because labor division usually results in higher labor productivity which, in turn, becomes observable in higher incomes, it could be argued that household incomes represent a suitable indicator of labor division. It is also assumed that rural and urban labor markets require different degrees of educational qualifications: high degrees of labor division and

Table 5.4 Summary of key variables characterizing rural and urban paths of adaptation in Bavaria

	Size of municipal district (number of inhabitants)				
	<5,000	5,000–20,000	20,000–100,000	100,000–500,000	>500,000
Number of observations	1,514	476	58	6	2
Inhabitants (mean)	2,307	9,072	34,305	146,562	929,425
Population density[1]	0.22	0.33	0.51	0.72	1.07
Dwelling technologies					
Apartment buildings[2]	0.10	0.23	0.46	0.62	0.75
Heating source[3]					
Firewood	0.08	0.04	0.03	0.01	0.01
Oil	0.72	0.52	0.29	0.16	0.09
Gas	0.14	0.38	0.61	0.72	0.68
District	–	0.01	0.03	0.09	0.21
Household size	2.41	2.24	2.01	1.83	1.82
Living space (m²/capita)	47.87	45.44	42.65	40.14	37.96
Cars (per capita)	0.60	0.59	0.54	0.50	0.45
Cultural preferences					
Percentage of property owners	0.72	0.60	0.46	0.33	0.28
Political preferences[4]					
CSU	0.48	0.45	0.41	0.36	0.32
SPD	0.14	0.16	0.17	0.20	0.21
Grüne	0.08	0.09	0.11	0.14	0.15
Family values[5]					
Weddings	48	48	50	48	40
Divorces	21	22	23	22	24
Illegitimate children	18	18	19	22	24
Religious affiliation[6]					
Catholic	0.66	0.60	0.49	0.44	0.34
Non-Catholic, non-Evangelical, or no religious affiliation	0.14	0.20	0.27	0.31	0.42
Modes of social reproduction					
Income[7]	0.95	1.03	1.09	1.06	1.20

Level of education (middle school or below)[8]					
Men	0.80	0.74	0.68	0.58	0.51
Women	0.80	0.75	0.71	0.63	0.56
Level of education (university degree)[8]					
Men	0.08	0.11	0.14	0.21	0.21
Women	0.04	0.06	0.08	0.14	0.15
Employment rate[9]					
Men, full-time	0.60	0.57	0.53	0.52	0.50
Men, part-time	0.02	0.03	0.03	0.04	0.04
Women, full-time	0.30	0.31	0.33	0.32	0.36
Women, part-time	0.39	0.38	0.37	0.36	0.30
Children[10]	0.16	0.15	0.14	0.13	0.12
CO_2 emissions per capita (t/year)					
Mobility	1.49	1.44	1.32	1.23	1.09
Housing	3.49	3.35	3.28	3.08	2.99
Mobility and housing	4.98	4.80	4.60	4.31	4.09
Consumption	5.15	5.60	5.92	5.49	6.17
Mobility, housing, and consumption	10.14	10.40	10.51	9.80	10.26

Source: compiled by the author. Bavarian municipality sample.

1 Inhabitants per 100 m² of constructed area.

2 Expressed as percental amount of living space in apartment buildings (in relation to the total amount of living space in the municipality).

3 Expressed as percental amount of living space built after 1981 heated by the respective fuel type (in relation to the total amount of living space in the municipality).

4 Expressed as percental amount of votes for the respective party (in relation to all counted votes, federal election 2009, second vote). Separate analysis for the last 60 years (beginning 1949) shows that the differences between rural and urban areas have increasingly converged over the course of time. CDU: Christian Democratic Union, SPD: Social Democratic Party, Grüne: The Greens.

5 Expressed as total number per 1000 inhabitants (reference level: rural districts, mean of the period 2001–2010).

6 Expressed as percental amount of the respective affiliation (in relation to the total amount of all reported affiliations, including those inhabitants with no religious affiliation).

7 Expressed as 1000 €/month per capita.

8 Percentage of men and women of the respective educational level (in relation to the respective total of men and women who are employed and are subject to social insurance contributions). Please note that only those individuals are captured here who are subject to social insurance contributions – self-employed persons or mini-jobbers are not included.

9 Percentage of men and women working full- or part-time (in relation to the respective total of men and women of working age). Please note that the number of persons working full- or part-time only captures those who are subject to social insurance contributions, meaning that self-employed persons or mini-jobbers are not included.

10 Percentage of children 14 years old or younger (in relation to the total population).

specialization should go hand in hand with higher educational qualifications (and vice versa). Thus, educational qualifications (of men and women) can be used as another indicator of labor division, too. Eventually, employment rates (of men and women) and the number of children complete the picture. For rural areas, it is expected that comparably low incomes coincide with high male employment rates, low female employment rates, and a high number of children. In urban areas, the ratio between male and female employment rates should be much more balanced, meaning that social reproduction is not achieved by (predominantly female) "do it yourself work", but is rather accomplished via paid labor and market exchange.[14]

Last but not least, working hypothesis 3 holds that rural and urban modes of life are accompanied by qualitative and quantitative differences regarding their metabolic character. Subchapter 5.1 suggested that urban areas are energetically more efficient than rural ones. In contrast, this subchapter assumes that the opposite becomes true when higher incomes in urban areas are taken into account, which was neglected in the previous approach.

Energy consumption and CO_2 emissions from mobility and housing were estimated using the regression equations for mobility and housing as presented in the models of subchapter 5.1. Using the standard regression equation

$$y = x_1 \times \beta_1 + x_2 \times \beta_2 + \dots + x_n \times \beta_n + con,$$

β-values were taken from the models presented in Table 5.2 whereas x-values were fed in from the municipality data described above. Based on this procedure, it is possible to describe the spatial distribution of CO_2 emissions resulting from mobility and housing in Bavarian municipalities (cf. Figure 5.8). As the determinants of rural and urban CO_2 emissions have already been scrutinized in subchapter 5.1, a descriptive analysis should suffice here. Besides, it does not make sense to explain CO_2 emissions by independent variables that have already been used as x-values in the estimation of the CO_2 emissions of Bavarian municipalities.

Concerning income-triggered CO_2 emissions, the CO_2 calculator as provided by the Federal Environmental Agency (UBA 2013) was utilized. Overall CO_2 emissions for the average German citizen were estimated at approximately 11 t CO_2 per year, of which roughly 5.58 t CO_2 are related to nutrition, aviation, public transportation, as well as all other forms of consumption (except for energy expenditures for private transport and housing). In order to estimate Bavarian income-triggered CO_2 emissions, the nationwide average of energy expenditure for private transportation and housing was subtracted from the nationwide average of household incomes. Next, the resulting nationwide residual incomes were divided by 5.58 t CO_2/capita, showing that each residual Euro – on average – results in 0.524 kg CO_2/capita in Bavaria.[15] The result was then multiplied by Bavarian municipality-specific residual incomes. In this way, income-triggered CO_2 emissions were estimated (cf. Figure 5.9). Totaling these different estimates of CO_2 emissions resulting from mobility, housing, and consumption finally resulted in the dependent variable that will be used to describe the metabolic properties of rural and urban modes of life as indicated in working hypothesis 3.

Admittedly, these rough procedures of estimating CO_2 emissions are not as accurate as one would wish them to be and could be rightly criticized for a variety of reasons. Especially with regard to income-triggered CO_2 emissions, critics could fault the resulting variable "CO_2 consumption" as not revealing anything about the spatial distribution of carbon emissions, but only about differences in rural and urban incomes. Therefore, it does not make sense here to run linear regression models to identify rural and urban determinants of income-triggered CO_2 emissions – such a procedure would only explain income differentials between rural and urban paths of adaptation. Moreover, the approach used to estimate income-triggered CO_2 emissions neglects possible differences in rural and urban consumption patterns (or savings behavior) but proceeds as if social actors in rural and urban paths of adaptation consumed comparable baskets – undoubtedly a disputable assumption (Häußermann/ Siebel 1996, Hertwich 2005). Given these pitfalls, what would be the alternatives?

Many Bavarian municipalities have recently started estimating their own climate balances. However, it is not advisable to use these data. Many municipalities developing their own methods and standards confuse consumption and production perspectives or – depending on political calculation – more or less obviously try to improve their climate balance by "fine-tuning" their local emission factors in rather intransparent ways. Comparing rural and urban metabolic characteristics on such a data basis is practically impossible. Thus, and even though the estimation techniques presented above are without doubt disputable, they appear to be much more reliable and transparent when compared with these alternatives. The estimation techniques used here have the advantage that the ß-coefficients, the x-values, and the CO_2 calculator are based on well-documented data sources such as the German Institute for Economic Research (DIW), the Federal Statistical Office, or the UBA. Besides, and in the light of the fact that the estimated Bavarian average of income-triggered CO_2 emissions amounts to 5.28 t CO_2/capita (as compared with the nationwide average of 5.58 t CO_2/capita), the estimation technique sketched out above seems to yield quite reasonable results. Given these pros and cons, income-triggered CO_2 emissions will be used here only for illustration. In view of the long-lasting scientific debate about climate change, rebound effects, and the huge challenge of decarbonizing the economy, the question is why ecological research has not yet come up with more reliable estimation techniques (Binswanger 2001, Jackson 2009). In this context, life cycle approaches to sustainable consumption as reviewed, for example, by Hertwich (2005) present a highly promising field of future research.

5.2.3 Detecting and visualizing spatial matches between value orientations, modes of social reproduction, and dwelling technologies

Before presenting the analytical findings, a few annotations should be made on the chosen analytical approach. Subchapter 5.1 identified the theoretically expected spatial patterns of dwelling technologies and, with the help of simple linear regression models, detected the main triggers of energy consumption and related CO_2 emissions within rural and urban paths of adaptation. Based on these findings, the present subchapter particularly asks whether modes of social reproduction and cultural preferences can also be operationalized, and whether rural and urban dwelling technologies are accompanied by cultural preferences and modes of social reproduction in the theoretically expected way. Given these research questions, a mainly descriptive analytical approach should suffice. Thereby, Cronbach's alpha (α) as well as bivariate correlations (Pearson's correlation coefficient, r) are used in order to check whether the selected variables measuring dwelling technologies or cultural preferences are reliable and whether the theoretically assumed correlations between these variables within rural and urban paths of adaptation actually exist. Unfortunately, and due to the small numbers involved (especially with regard to municipalities with more than 100,000 inhabitants), the comparison of means could not be applied.

Starting with the first working hypothesis dealing with the correspondence between different dwelling technologies and cultural preferences, Table 5.4 yields the theoretically expected results: increasing degrees of urbanization and physical compactness (as measured by the number of inhabitants per 100 m²) go hand in hand with a steady decline in grid-unconnected heating sources such as firewood or oil. Grid-connected energy carriers such as gas or district heating, as well as the percentage of living space within apartment buildings, significantly increases with growing urbanization (cf. Figure 5.1). In addition, household size, living space per capita, and the number of cars per capita steadily decline with increasing degrees of physical compactness.

With a Cronbach's alpha above 0.80 ($\alpha = 0.81$), the selected variables measuring the construct "dwelling technologies" appear to be highly reliable. This is also confirmed when correlation coefficients are investigated. To give an example, increasing levels of population density are negatively correlated with the amount of living space heated by oil ($r = -0.41$; $p < 0.001$) and positively correlated with the area of living space heated by gas ($r = 0.46$; $p < 0.001$). Likewise, increasing levels of population density are negatively correlated with household size ($r = -0.55$; $p < 0.001$). In sum, the spatial distribution of dwelling technologies confirms the theoretically expected correspondences between physical compactness, grid-connected energy carriers, and household size as identified in subchapter 5.1. But do dwelling technologies and cultural preferences also correspond in the theoretically expected way?

To begin with, it is assumed that the construct "cultural preferences" can be captured by the amount of real-estate owners, by political preferences, by specific family values, and by religious affiliation. With a Cronbach's alpha close to 0.80 ($\alpha = 0.79$), these variables show a high internal consistency and thus represent reliable indicators measuring cultural preferences. As shown in Figure 5.2 and as expected theoretically, rural areas are characterized by a large share of real-estate owners (taken as an indicator of possessive individualism), by rather right-wing and conservative political orientations (CSU), and by traditional family values

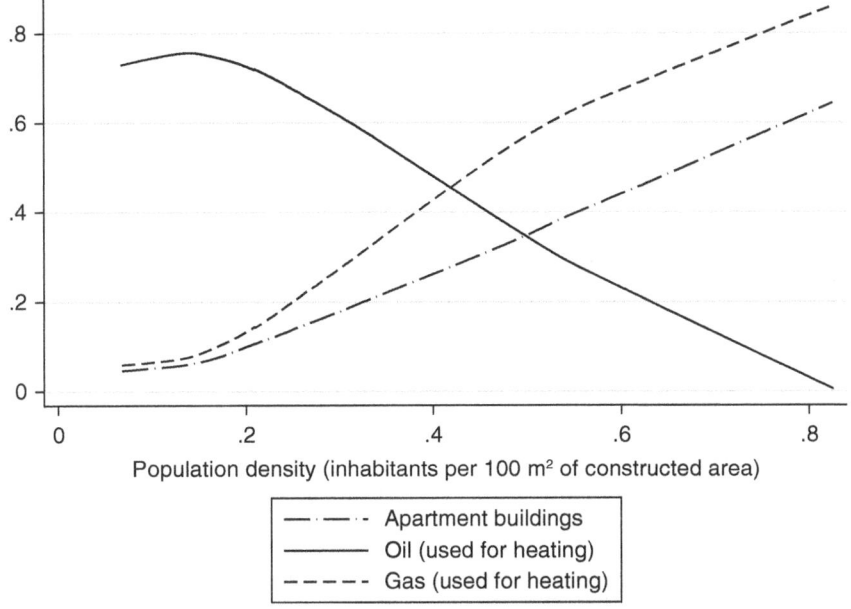

Figure 5.1 Correspondences between population density and dwelling technologies
Source: compiled by the author. Bavarian municipality sample.
Note: Lowess smoothing was applied here.
Specification of variables: cf. Table 5.4.

(indicated by comparably high numbers of weddings combined with lower numbers of divorces and illegitimate children). Taking the large proportions of CSU sympathizers and Catholics, one could conclude that rural paths of adaptation are actually marked by more coherent cultural preferences and value orientations.

Urban paths of adaptation, by contrast, appear to be culturally more heterogeneous, as indicated, for example, by both the more balanced shares of party affiliation and the large number of individuals not affiliated to either of the two main persuasions or having no religious orientation at all. This trend is also accompanied by a significantly lower amount of real-estate owners.

Regarding the correspondence between dwelling technologies and cultural preferences, a Cronbach's alpha above 0.80 ($\alpha = 0.87$) indicates that the respective variables describing dwelling technologies and cultural preferences are characterized by a high internal consistency. This could be interpreted as further evidence indicating that the selected variables are well suited to capturing the interplay between dwelling technologies and cultural preferences in rural and urban paths of adaptation. As illustrated in Table 5.4, the variables describing dwelling technologies and cultural preferences actually coincide in the theoretically expected spatial ways. With increasing degrees of urbanization and population density, the proportions of CSU voters, owners, and Catholics as well as the number

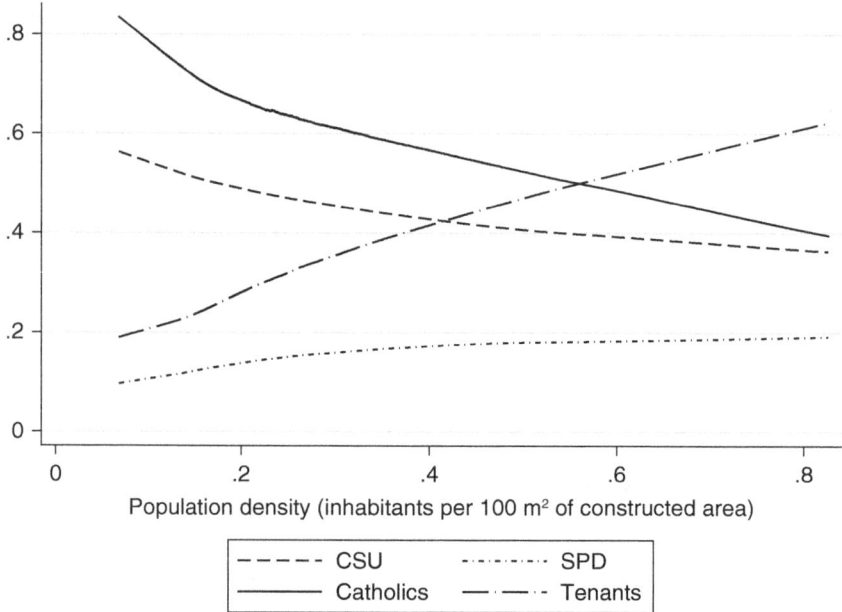

Figure 5.2 Correspondences between population density and cultural preferences
Source: compiled by the author. Bavarian municipality sample.
Note: Lowess smoothing was applied here.
Specification of variables: cf. Table 5.4.

of unconnected dwelling technologies more or less show a parallel decline (cf. Figure 5.3). The opposite is true when analyzing the number of tenants, the number of connected dwelling technologies, and the proportions of SPD voters and individuals not affiliated to either of the two main persuasions or having no religious orientation at all (cf. Figure 5.4). However, this does not mean that all CSU voters are Catholics who live in detached houses heated by oil – this would be an ecological fallacy (Diekmann 2008). These findings simply underline the theoretical assumption according to which specific correspondences between dwelling technologies and cultural preferences are more likely to be found in rural than in urban paths of adaptation (and vice versa).

An examination of the bivariate correlation among dwelling technologies, cultural preferences, and political orientations on the one side and building type and heating source on the other demonstrates especially clear results: large proportions of CSU voters are negatively correlated with physically compact building types such as apartment buildings ($r = -0.37$; $p < 0.001$) and grid-connected heating technologies such as gas ($r = -0.43$; $p < 0.001$) or district heating ($r = -0.15$; $p < 0.001$). The same holds true for Catholic religious affiliation. In contrast, social-democratic political preferences (as well as non-Catholic, non-Evangelical/ non-Protestant, or no religious affiliation at all) are positively correlated with physically compact forms of dwelling and related grid-dependent fuel types.

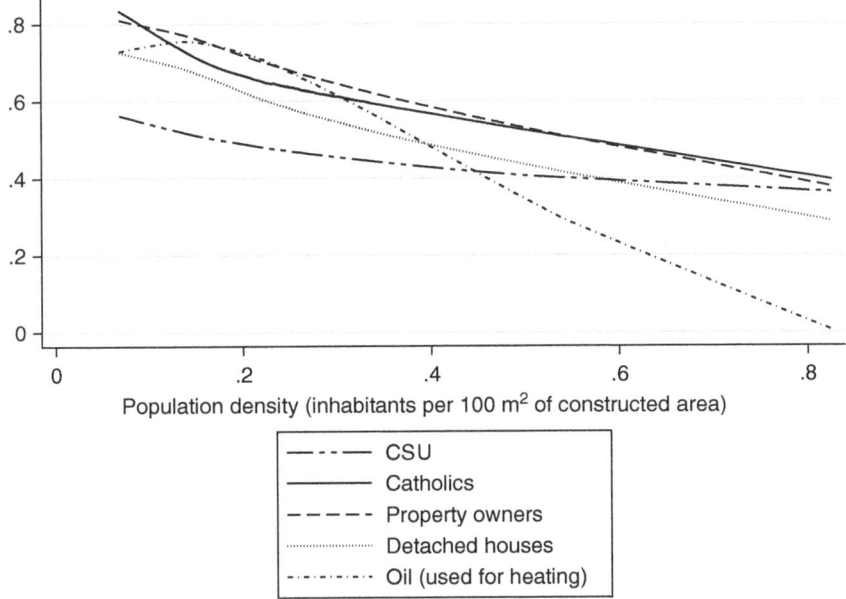

```
— · · — CSU
———— Catholics
— — — — Property owners
· · · · · · · · · Detached houses
— · · — · · — · · Oil (used for heating)
```

Figure 5.3 Correspondences among population density, dwelling technologies, and traditional-conservative cultural preferences
Source: compiled by the author. Bavarian municipality sample.
Note: Lowess smoothing was applied here.
Specification of variables: cf. Table 5.4.

In sum, these findings confirm the first working hypothesis according to which urban dwelling technologies should coincide with non-conservative and social-democratic value orientations, whereas rural dwelling technologies should be accompanied by conservative-individualistic value orientations.

The second working hypothesis states that rural and urban paths of adaptation should not only show distinct matches between dwelling technologies and cultural preferences, but should also be accompanied by appropriate modes of social reproduction in the broadest sense. Table 5.4 shows that the construct "modes of social reproduction" is measured by per capita incomes, male and female levels of education and employment rates, as well as by the proportion of children (14 years old or younger). With regard to educational levels and employment rates, note that only those individuals subject to social insurance contributions are captured here – self-employed persons or mini-jobbers are not included. Thus, the following discussion should be read with some reservations. Despite this, a Cronbach's alpha above 0.80 ($\alpha = 0.82$) indicates that the selected variables show a high internal consistency which, in turn, suggests that the theoretical considerations which inspired the selection of variables measuring modes of social reproduction are practicable. As expected, rural modes of social reproduction are actually marked by comparably lower incomes (and lower educational levels,

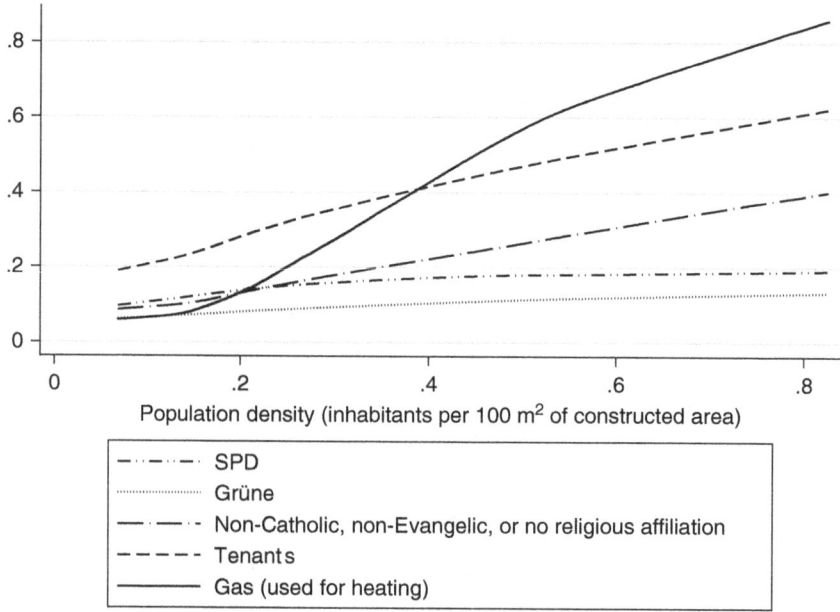

Figure 5.4 Correspondences among population density, technologies, and alternative or left-wing cultural preferences
Source: compiled by the author. Bavarian municipality sample.
Note: Lowess smoothing was applied here.
Specification of variables: cf. Table 5.4.

respectively) which, in turn, are accompanied by higher proportions of (unpaid) female "do it yourself work" and slightly more children. In sum, this constellation reveals a lot about male and female role models in rural and urban paths of adaptation and punctuates the conservative-traditional mix of possessive individualism, CSU voters, and Catholic affiliation as diagnosed above for rural areas (with men as the main breadwinners).

To provide further evidence from the respective bivariate correlation matrix, the share of CSU voters positively correlates with male and female educational levels of middle school or below (men: $r = 0.38$; $p < 0.001$; women: $r = 0.32$; $p < 0.001$) and negatively correlates with male and female university degrees (men: $r = -0.36$; $p < 0.001$; women: $r = -0.31$; $p < 0.001$). Furthermore, the proportion of CSU voters positively correlates with male full-time employment and negatively correlates with female full- and part-time employment as well as with male part-time employment. Thus, it should come as no surprise that physically connected dwelling technologies positively correlate with the proportion of university degrees, whereas comparatively lower educational levels positively correlate with grid-unconnected heating sources and detached housing (cf. Figure 5.5).

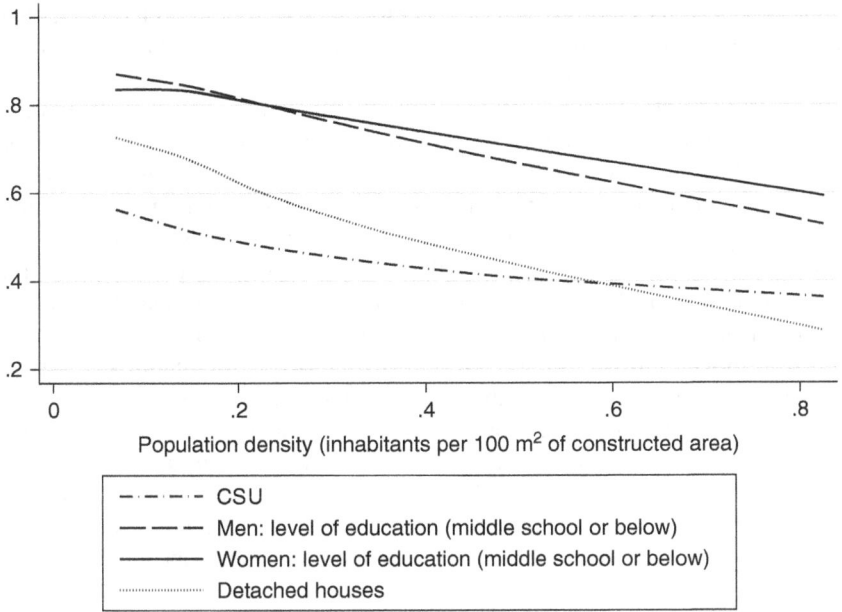

Figure 5.5 Correspondences among population density, traditional-conservative cultural preferences, and educational levels (middle school or below)
Source: compiled by the author. Bavarian municipality sample.
Note: Lowess smoothing was applied here.
Specification of variables: cf. Table 5.4.

With increasing degrees of urbanization, the proportions of men and women working full-time converge to a certain degree (at least with those who are subject to social insurance contributions), indicating that social reproduction is accomplished via paid labor and market exchange (instead of "do it yourself work") in urban paths of adaptation. As paid labor and market exchange amongst strangers heavily depend on third-party enforcement (rather than cultural homogeneity, North 1990), it was theoretically expected that urban modes of social reproduction would tolerate higher degrees of cultural heterogeneity and should be more open-minded toward value orientations emphasizing and practicing anonymized solidarity (for example, via welfare arrangements). In this context, it could be argued here that the comparatively larger shares of SPD and Grüne voters, as well as the substantial numbers who are non-Catholic, non-Evangelical/non-Protestant, or have no religious affiliation at all, point to the anticipated direction. Based on these findings, the second working hypothesis can also be confirmed.

Taken together, it was argued that rural and urban paths of adaptation can be measured by dwelling technologies, cultural preferences, and modes of social reproduction. Calculating an overall Cronbach's alpha higher than 0.90 ($\alpha = 0.92$)

for all variables measuring dwelling technologies, cultural preferences, and modes of social reproduction could be seen as an additional clue that the selected variables actually coincide in the theoretically expected ways.[16]

Concerning the third working hypothesis, it was assumed that the combined effects of dwelling technologies, cultural preferences, and different modes of social reproduction would result in distinct patterns of CO_2 emissions in rural and urban paths of adaptation. Apart from the fact that total CO_2 emissions stemming from mobility, housing, and consumption are more or less the same in rural and urban paths of adaptation, Table 5.4 yields the theoretically expected results (at least at a first glance): CO_2 emissions from mobility and housing steadily decrease with increasing degrees of urbanization. As described in subchapter 5.1, these differences in CO_2 emissions can largely be explained by household size, living space per capita, building type (high or low physical compactness), and the availability of grid-connected energy carriers with comparatively low carbon content. Here again, it becomes apparent that social and physical concentration counteract each other: whereas household size declines with increasing degrees of urbanization, the share of physically connected dwelling technologies increases evermore. Thus, the third working hypothesis (as well as the findings in subchapter 5.1) are also confirmed.

Generally, the same should also hold true for income-triggered carbon emissions. Although it was argued that the estimation of income-triggered carbon emissions proceeded somewhat inexactly, this presents a precise illustration of incomes. As the relevant literature agrees that carbon emissions are likely to increase with rising incomes (Hertwich 2005), income-triggered carbon emissions in urban areas should be higher than those in rural settlement structures. However, the following maps (Figures 5.6–5.10) showing the spatial distribution of CO_2 emissions in Bavarian municipalities suggest that different degrees of physical compactness and related (infra-)structural peculiarities present necessary, yet insufficient, theoretical arguments to explain the variation in carbon emissions in rural and urban paths of adaptation.

Examining agglomerations such as Munich, Augsburg, Nuremberg, and their surrounding municipalities, Figure 5.6 demonstrates that significant differences exist in CO_2 emissions stemming from individual transportation. However, these differences do not follow precisely the theoretically expected spatial distribution. Apparently, the level of carbon emissions stemming from mobility is not determined only by the number of inhabitants and related (infra-)structural properties, but probably also by the spatial location of communities (for example, in terms of accessibility (i.e. remoteness vs. centrality or low or high market proximity, respectively). How else should one explain the significant differences in carbon emissions despite comparable district size (for example, between Munich exurbs and municipalities located at the German–Czech border in the east?). Here, and with regard to different modes of social reproduction, it could be argued that remote communities are more self-sufficient than those close to urban centers, meaning that variations in carbon emissions stemming from individual transportation can be explained by the fact that municipalities close to agglomerations

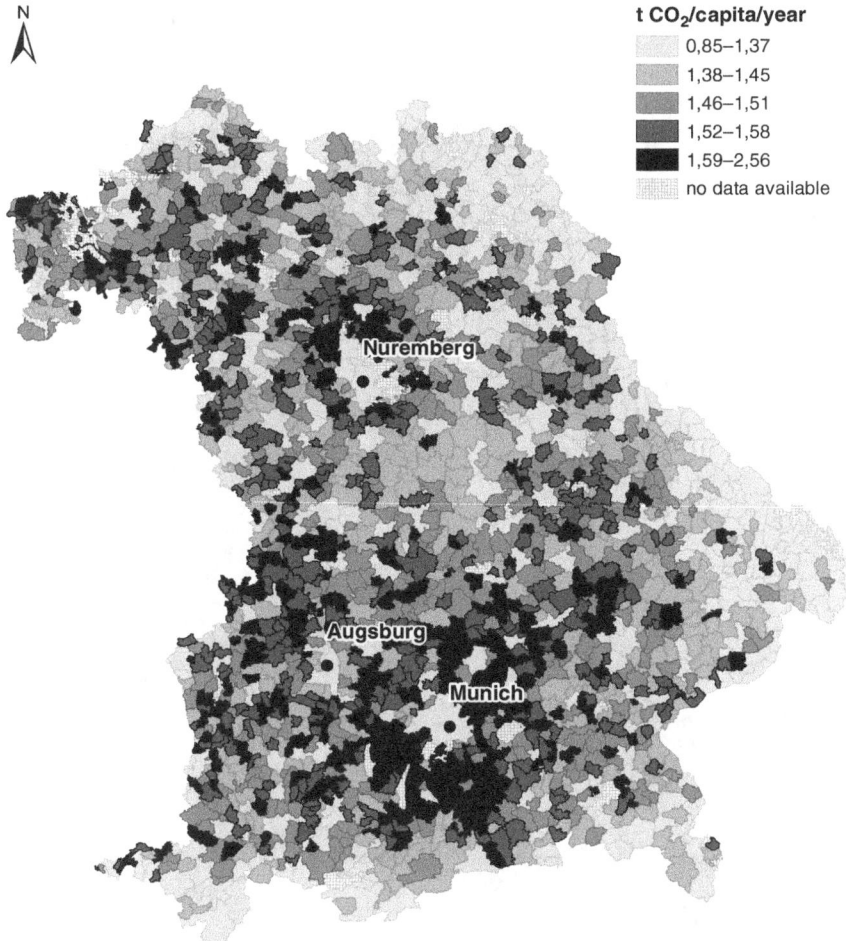

Figure 5.6 Spatial distribution of CO_2 emissions resulting from individual transportation (t CO_2/capita per year)
Source: compiled by the author. Bavarian municipality sample.

have a larger share of commuters. Moreover, municipalities in the immediate vicinity of agglomerations also feature comparatively higher incomes which, in turn, translate into more and larger cars with more powerful (and more fuel-intensive) engines.

The spatial distribution of CO_2 emissions resulting from housing operations yields comparative results (cf. Figure 5.7): remote municipalities – on average – produce much higher carbon emissions than communities located in or close to agglomerations. Primarily, this effect can be explained by the fact that living space per capita in peripheral regions is significantly above average (due to lower rents and land prices). In contrast, and regarding the exurbs located in the southwest of

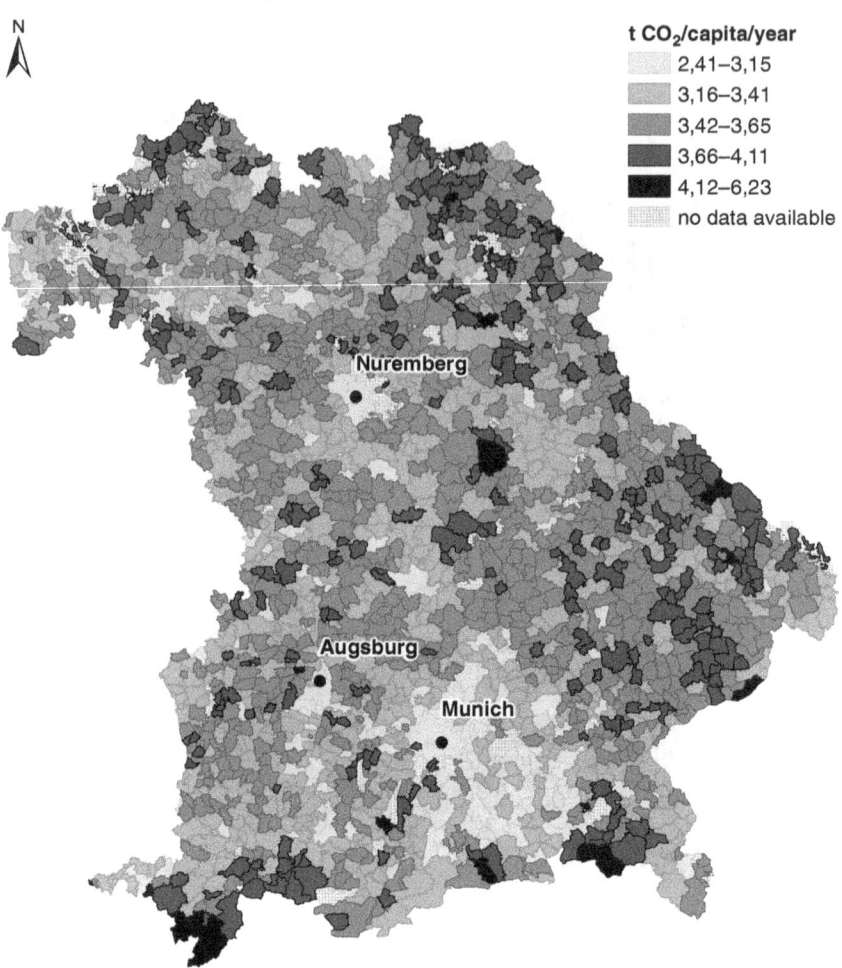

Figure 5.7 Spatial distribution of CO_2 emissions resulting from building operations such as heating or hot water supply (t CO_2/capita per year)
Source: compiled by the author. Bavarian municipality sample.

Munich (for example, Starnberg, Berg, or Bernried), above average living space per capita is mainly due to high incomes.

Taken together, and as illustrated in Figure 5.8, the spatial distribution of CO_2 emissions caused by mobility and housing appears to be quite patchy. Even though CO_2 emissions in urban agglomerations such as Munich, Nuremberg, or Augsburg – on average – are significantly lower than those in their high-income exurbs, the widely distributed light gray patches in comparatively remote areas suggest that different degrees of urbanization and related (infra-)structural peculiarities present necessary yet insufficient theoretical arguments to explain different carbon emissions in rural and urban paths of adaptation. In this context, it could be argued that carbon emissions stemming from mobility and housing (and related dwelling technologies) are determined by physical compactness, whereas income-triggered CO_2 emissions vary with centrality and remoteness (high or low labor division and corresponding cultural preferences). Despite the fact that physical compactness and accessibility overlap spatially, this differentiation – as compared with the hitherto one-dimensional approach based on high or low physical compactness – should result in a more fine-grained picture, shedding more light on the spatial correspondence of dwelling technologies, modes of life, and related patterns in resource consumption.

This assumption is supported by Figure 5.9. Here, it can be observed that incomes and related income-triggered CO_2 emissions clearly follow the distinction between centrality and remoteness. Whereas Table 5.4 seems to imply that there is no noteworthy difference in average total carbon emissions (stemming from mobility, housing, and consumption), Figure 5.10 reveals remarkable spatial variations.

As suggested by Gill and Schubert (forthcoming), discerning municipalities by a two-dimensional grid (with high or low physical compactness as the *x*-axis and accessibility indicating market proximity as the *y*-axis) will deliver further valuable insights. Applying this twofold typology, the first results suggest the following distribution of total CO_2 emissions per year (t CO_2/capita): 10.07 in rural villages (low physical compactness, poor accessibility), 9.83 in provincial towns (high physical compactness, poor accessibility), 11.36 in suburbs (low physical compactness, good accessibility), and 10.96 in urban agglomerations (high physical compactness, good accessibility). As shown by these figures, the suburban mode of life – due to low physical compactness and high incomes – is accompanied by the highest carbon emissions. Moreover, and in accordance with Bettencourt et al. (2007), the negligible differences in carbon emissions contradict the popular assumption according to which the ecological crisis could be solved by increasing levels of urbanization (Glaeser/Kahn 2010, Glaeser 2011).

5.2.4 Summary

The advancement of eco-cultural coevolution claims that fit matches between physical nature, technologies, institutions, and cultural preferences emerge over the course of time. By conceptualizing rural and urban settlement structures and related modes of life as distinct paths of eco-cultural adaptation, this idea has undergone an initial empirical test in the present chapter. Subchapter 5.1 revealed that it is possible to operationalize the technological dimension of eco-cultural matches and that rural and urban settlement structures differ from each other in the theoretically expected way (at least with regard to dwelling technologies). Based on these findings, subchapter 5.2 pursued two central research questions: is it also possible to operationalize cultural preferences and modes of social reproduction and – if so – do they spatially correspond with dwelling technologies in the theoretically expected way?

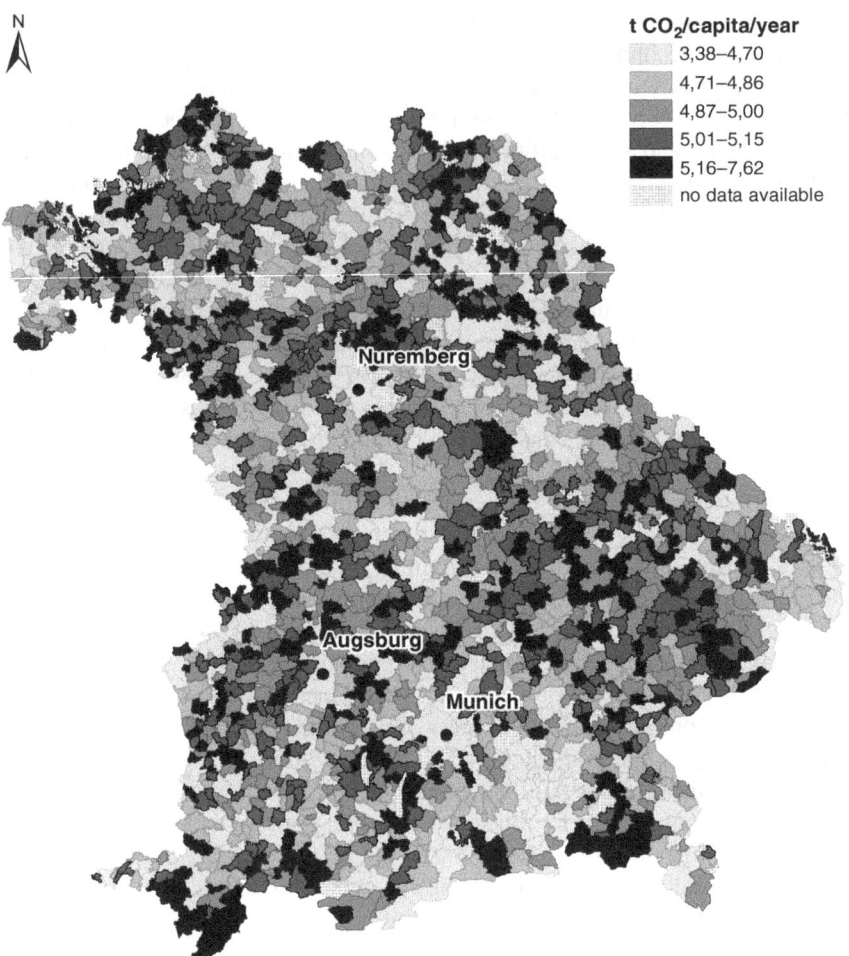

Figure 5.8 Spatial distribution of CO_2 emissions resulting from individual transportation and building operations (t CO_2/capita per year)
Source: compiled by the author. Bavarian municipality sample.

Altogether, it has transpired that the answers to both questions are affirmative. The present subchapter demonstrated a manageable solution to the operationalization of cultural preferences and modes of life. Apart from some minor inconsistencies (for example, with regard to educational levels or employment rates, which were only available for those who were subject to social insurance contributions, meaning that self-employed persons or mini-jobbers were not captured here), and the admittedly rough estimates of income-triggered CO_2 emissions, the operationalization has produced convincing results. This conclusion is also emphasized by the reported Cronbach's alpha values, which attest that the operationalization procedure is marked by strong internal consistency among the selected variables

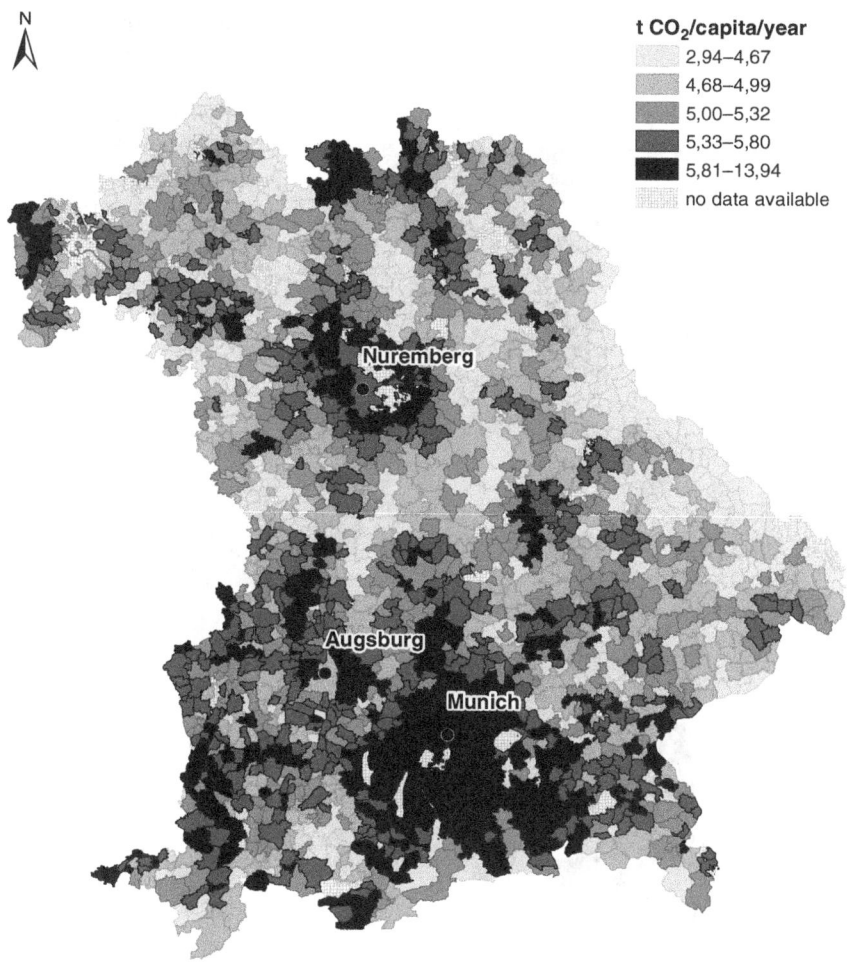

Figure 5.9 Spatial distribution of income-triggered CO_2 emissions (t CO_2/capita per year) Source: compiled by the author. Bavarian municipality sample.

(which, in turn, indicates that the underlying theoretical considerations go in the right direction). Dwelling technologies, cultural preferences, and modes of social reproduction coincide in the theoretically assumed way.

In rural paths of adaptation, spatial dispersion and physically unconnected technologies correspond with possessive-individualistic and conservative-traditionalistic family values and political preferences. In urban paths of adaptation, in contrast, physical compactness and connected technologies go hand in hand with social-democratic political preferences and a comparatively broader degree of cultural heterogeneity and related family and life plans. The analytical results as summarized in Table 5.4 also corroborate the findings of

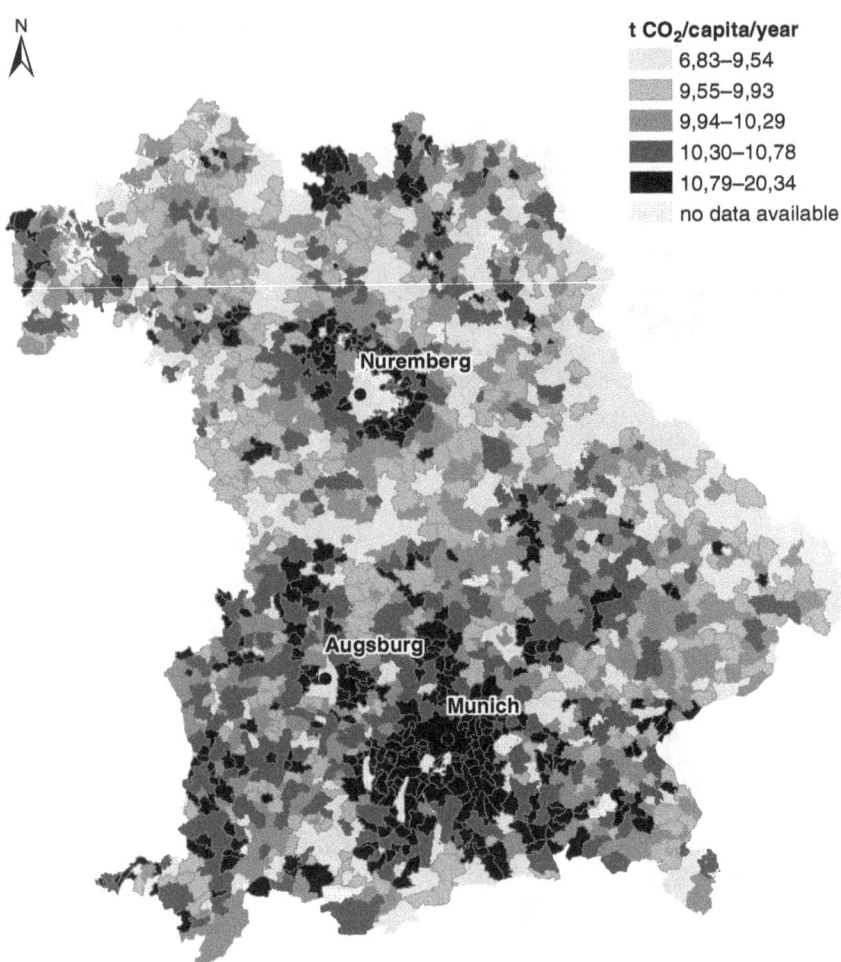

Figure 5.10 Spatial distribution of total CO_2 emissions resulting from individual transportation, building operations, and consumption (t CO_2/capita per year)
Source: compiled by the author. Bavarian municipality sample.

subchapter 5.1 where carbon emissions were assessed: CO_2 emissions caused by individual transportation and building operations are significantly higher in rural than in urban areas, emissions decline with increasing degrees of urbanization, and the reverse effects of physical and social concentration counteracting each other are also observable.

Concerning the overall CO_2 balances of rural and urban paths of adaptation, it was also assumed that rural paths of adaptation should score considerably better when income-triggered CO_2 emissions were taken into account (Hertwich 2005, Bettencourt et al. 2007). Although this question could not be finally answered

here, the observable differences in carbon emissions (caused by individual transport and housing) between rural and urban areas should more or less converge
when income differentials are considered: the data presented reveal that urban
paths of adaptation actually show significantly higher incomes which – according to the relevant literature – should translate into higher carbon emissions.
However, more research regarding spatial differences in consumption patterns
and the measurability of income-driven carbon emissions has to be done before
spatial differences in consumption-based CO_2 emissions can be reliably described
in absolute terms.

Staying with the subject of further research, the graphical representation of
carbon emissions (cf. Figures 5.6–5.10) questioned the findings of Table 5.4 to
a certain degree. Whereas Table 5.4 seems to suggest that the selected variables
change more or less linearly with the number of inhabitants, the maps showing the spatial distribution of carbon emissions suggest that a one-dimensional
description of either rural or urban paths of adaptation does not suffice. Here,
the introduction of a second dimension accounting for centrality and remoteness (used as a proxy for market proximity) should yield further insights (Gill/
Schubert, forthcoming). However, a two-dimensional approach does not refute
the general findings of this subchapter, but results in some more pronounced
spatial effects, especially with regard to small municipalities lying in the
wealthy exurbs of urban agglomerations, where rural dwelling technologies are
likely to converge with urban modes of social reproduction). Against the background of these results, three central assumptions of the present main chapter
can be confirmed: the approach of eco-cultural adaptation is operationalizable
and yields the theoretically expected results. Therefore, it is also marked by a
high predictive capacity.

Regarding the concept of spatial fractionalization (or self-similarities), one
question has not yet been answered: is it true that the same correspondences
between dwelling technologies, cultural preferences, and modes of social reproduction can also be observed within urban eco-cultural habitats (i.e. on a smaller
scale?) On the basis of self-collected data from Munich and Bolzano, this question will briefly be dealt with next.

5.3 Self-similarities: intra-urban patterns of density and dispersion in Munich and Bolzano

In simple terms, the basic idea of fractionalization as elaborated, for example,
by Mandelbrot (1967, 1987) is that self-similar geometric patterns can be found
on different analytical scales. Recently, the concept of fractionalization has been
applied more and more to gain deeper understanding of cities' socio-structural
properties (Bettencourt et al. 2007, Batty 2011, 2013, Bettencourt 2013).
Inspired by this relatively new field of research, and based on the findings of subchapters 5.1 and 5.2, the present subchapter will examine whether similar technological, cultural, and economic differences characterizing rural and urban paths of
adaptation can be observed within urban settlement structures, too, for example,

regarding the correspondence among building types (for example, detached or semi-detached houses vs. apartment buildings), political preferences, and family values (for example, traditional-conservative vs. social-democratic preferences).

5.3.1 Working hypotheses: patterns of rurality and urbanity can also be found within the city

The present subchapter adopts the working hypotheses discussed in subchapter 5.2, with no major modifications. Whereas rural and urban dwelling technologies were mainly captured by physical compactness (in terms of district size or inhabitants per 100 square meters of constructed area) in subchapter 5.2, the present subchapter distinguishes between different intra-urban building types (ranging from detached houses to apartment buildings with six or more stories) to account for varying degrees of physical concentration. As will shortly become apparent, the adopted working hypotheses partly collide with our everyday experience regarding intra-urban forms of dwelling and related cultural preferences as well as modes of social reproduction. However, and in order to either verify or disprove the central idea of fractionalization as recapitulated above, it is important to pursue comparable working assumptions.

First, and concerning the correspondence between dwelling technologies and cultural preferences, it is assumed that physically connected dwelling technologies (for example, apartment buildings) correspond with social-democratic value orientations and a comparatively broader range of family and life plans. Unconnected dwelling technologies (for example, detached housing) should be accompanied rather by traditional-conservative value orientations and related life plans.

Second, and considering the interplay among dwelling technologies, cultural preferences, and observable modes of social reproduction, the previous chapter showed that rural paths of adaptation (i.e. physically unconnected dwelling technologies) coincide with comparably lower degrees of labor division and presumably more "do it yourself work" (as measured by educational level, form of employment, income, and number of children) whereas urban adaptational paths correspond with high degrees of labor division. Whether similar constellations can also be observed within cities is an open question – if so, occupiers of (semi-) detached houses should feature, for example, lower educational levels, lower incomes, and more children than those living in apartment buildings.

Third, per capita CO_2 emissions stemming from individual transportation and building operations should decline with increasing degrees of physical connectedness. In contrast, and as demonstrated in the previous chapter, household incomes (and related consumption-driven CO_2 emissions) should increase with rising levels of physical connectedness. Again, it is questionable whether physically connected dwelling technologies really coincide with higher incomes within urban paths of adaptation.

In general, it is assumed that all effects related to allometric rules (for example, SA:V ratio) or economies of scale (for example, decreasing energy expenditures with increasing household size) can also be found within urban paths of adaptation, regardless of whether Munich or Bolzano is the focus. Concerning the matches among dwelling technologies, cultural preferences, and modes of social reproduction as identified in the previous chapter, more ambiguous results are expected. To give one example, due to higher rents and land prices within urban settlement structures, well-educated and high-income actors are more likely to live in detached houses in urban areas whereas less educated persons with lower incomes can only afford apartment buildings.

5.3.2 Examining the relationship between socio-structural characteristics and energy consumption in urban agglomerations

The data presented here were recorded during a triennial research project funded by the German Ministry of Education and Research.[17] One of the central research questions of this project was the analysis of the relationship between socio-structural peculiarities and energy consumption in urban agglomerations in Bavaria (Germany) and the autonomous province of South Tirol (Northern Italy). As there are no comparable data problematizing the interplay of socio-structural characteristics and energy consumption on the household level in urban areas (even less so when two cities in different countries are concerned), the respective data had to be collected by the project team in 2012 via questionnaires.

A 12-page questionnaire of 55 questions was constructed. The questionnaire covered various subjects ranging from dwelling technologies, energy expenditures, and individual transportation (modal split) to socio-demographic aspects such as household size, household income, marital status, political and religious affiliation, and number of children. Where possible, respective item sets and questions were adopted from the GSOEP waves of 1998 and 2003 as described in subchapter 5.1. In doing so, and due to the fact that the GSOEP is well known for its conscientious design of questions and item sets, not only data comparability but also crucial quality criteria were ensured.

In Munich, the sample was drawn via a random route in a preselected city region spreading out from the city center to the south. Based on municipal data, the respective city region was selected due to its heterogeneous stock of buildings and a broad range of household types and incomes. In sum, 3300 questionnaires were sent to the selected households (including a stamped, addressed envelope for the return of the questionnaire). After one week, a reminder asking for participation was mailed to all households. A return rate of 32% ($N = 1060$) was achieved.

Bolzano is much smaller than Munich, so city-wide sampling could be realized. Before sampling, the questionnaire used in Munich was adapted to Bolzano, meaning that an Italian-language version was set up and some questions were revised (for example, with regard to educational achievements, political parties, or particularities regarding everyday language and local dialects). After elaborate pretests and a second revision round, 7000 questionnaires were distributed city-wide via postal mailing whereby one in seven households was contacted. Each household received one German and one Italian version as well as a stamped, addressed envelope for the return of the questionnaire. Regarding the distribution process, some irregularities were reported by survey participants. Although it was agreed that the questionnaires were to be delivered to one in seven households, a number of envelopes containing the questionnaires were found in the stairwell or among print advertising and not in individual letter-boxes. Therefore, the data collected in Bolzano cannot be described as a random sample in the strict sense. Despite these irregularities, a return rate of nearly 16% ($N = 1105$) was achieved.

Regarding the representativity of both samples, comparison by means of key socio-demographic variables such as gender, household size, number of children, and age with the available official statistics revealed significant deviations. Household size, number of children, and numbers of older and male participants were significantly above average whereas the numbers of single households and age groups younger than 30 were significantly below average in both samples. The general impression is that the well-educated middle class was especially attracted by the survey and its ecological topics such as energy consumption and carbon emissions (self-selection). Regarding heating technologies, building type, or living space per capita, official statistics are somewhat undifferentiated (highly aggregated), meaning that reliable representativity tests could not be conducted.

Initially, and analogous to subchapter 5.1, the data presented here were collected to learn more about the structural determinants of energy consumption and related carbon emissions in urban settlement structures. In this context, insufficient representativity is a somewhat negligible problem. However, as one searches for correspondences between dwelling technologies, cultural preferences, and modes of social reproduction, insufficient representativity results in major problems regarding the interpretation of crucial descriptive findings, as summarized in Table 5.5. Here, it cannot be said for sure whether the observable correspondences between dwelling technologies, cultural preferences, and modes of social reproduction are applicable to all of Munich and Bolzano or apply only to the respective sample. Therefore, census-like data as used in subchapter 5.2 would be more appropriate and reliable when questions regarding city-wide correspondences among dwelling technologies, cultural preferences, and modes of social reproduction occupy center stage. However, and because such data are not available, there is no other option but to critically reflect the observable correspondences summarized below.

Coming back to the first working hypothesis dealing with the correspondence between dwelling technologies and cultural preferences, households were asked to report the main energy carrier used for water and room heating as well as household size, living space, and the number of cars owned. Regarding cultural preferences, those polled gave an account of their ownership situation (owner or tenant), political preferences, marital status, and religious affiliation.

Concerning the second working hypothesis, modes of social reproduction can be measured by household income, educational level, type of employment, and the number of children living in the household.

Table 5.5 Munich and Bolzano: correspondences among dwelling technologies, cultural preferences, and modes of social reproduction

	Building type				
	Detached house	Row or semi-detached house	Detached two- or three-family house	Apartment building (up to five stories)	Apartment building (six stories or more)
Number of observations	77	89	139	1,531	283
Dwelling technologies					
Heating source[1]					
Firewood	0.31	0.11	0.14	0.03	0.01
Oil	0.18	0.16	0.13	0.09	0.09
Gas	0.43	0.64	0.55	0.57	0.55
District	0.01	0	0.04	0.09	0.12
Household size	3.05	2.61	2.48	2.17	2.26
Living space (m/capita)	59.84	53.05	56.97	45.74	43.56
Cars (per capita)	0.80	0.69	0.79	0.73	0.76
Cultural preferences					
Percentage of property owners	0.86	0.69	0.69	0.48	0.60
Percentage of tenants	0.13	0.31	0.30	0.51	0.40
Political preferences[2]					
Center-right	0.38	0.39	0.39	0.21	0.16
Center-left	0.13	0.14	0.12	0.19	0.28
Grüne	0.23	0.24	0.14	0.24	0.12
Marital status[3]					
Single	0.06	0.05	0.10	0.17	0.12
Unmarried with partner	0.06	0.13	0.12	0.21	0.14
Married	0.74	0.68	0.59	0.44	0.51
Divorced	0	0.06	0.07	0.06	0.07
Church attendance[4]					
Weekly	0.18	0.16	0.22	0.12	0.18
Monthly	0.21	0.15	0.17	0.10	0.13
Seldom	0.51	0.56	0.47	0.61	0.59
Never	0.08	0.13	0.13	0.15	0.10
Modes of social reproduction					
Age					
Men	64.07	59.39	57.84	55.20	60.93
Women	53.55	54.61	56.00	51.24	55.31
Retired					
Men	0.46	0.43	0.38	0.36	0.55
Women	0.20	0.38	0.26	0.26	0.33

Table 5.5 (cont.)

	Building type				
	Detached house	Row or semi-detached house	Detached two- or three-family house	Apartment building (up to five stories)	Apartment building (six stories or more)
Income[5,8]	1.49	1.77	2.04	2.27	2.46
Level of education (middle school or below)[6,8]					
Men	0.13	0.26	0.45	0.23	0.26
Women	0.22	0.34	0.51	0.34	0.38
Level of education (university degree)[6,8]					
Men	0.64	0.72	0.83	0.80	0.75
Women					
	0.88	0.74	0.55	0.75	0.73
Type of employment[7,8]					
Men, full-time	0.86	0.82	0.89	0.88	0.92
Men, part-time	0	0.07	0.04	0.04	0.04
Women, full-time	0.33	0.52	0.43	0.59	0.58
Women, part-time	0.42	0.30	0.38	0.23	0.28
Children[8]	1.49	1.71	1.45	1.15	1.24
Care of old or ill relatives[9]					
Yes	0.81	0.83	0.80	0.74	0.75
No	0.05	0.10	0.05	0.13	0.12
CO$_2$ emissions per capita (t/year)[10]					
Mobility	5.02	3.59	3.66	3.24	3.10
Housing	2.49	2.89	2.63	2.00	2.06
Mobility and housing	3.49	4.11	3.71	2.95	2.92
Consumption	1.81	1.51	2.17	2.11	2.23

Source: compiled by the author. Munich & Bolzano datasets.

[1] Relative frequencies of heating source in percent (within the respective building type). Because not all heating sources are reported here, column percentages do not exactly add up to 100.

[2] Relative frequencies of political preferences (within the respective building type). Because not all parties are reported here, column percentages do not add up to 100. Grüne: Greens.

[3] Relative frequencies of marital status (within the respective building type). Because not all categories are reported here, column percentages do not add up to 100.

[4] Relative frequencies of church attendance.

[5] Expressed in 1000 €/month per capita.

[6] Relative frequencies of men and women of the respective educational level (within the respective building type).

[7] Relative frequencies of the respective type of employment (within the respective building type).
[8] Here, only gainfully employed persons are included – otherwise, the large number of pensioners would bias the proportions between male and female full- and part-time employment and corresponding numbers of children.
[9] Here, people were asked whether they could rely on their families in case of care dependency (for example, due to illness or age-related frailty (relative frequencies)). Because the category "Maybe" is not reported here, column percentages do not add up to 100.
[10] As a result of different numbers of observations, the subtotals of CO_2 emissions resulting from mobility and housing do not add up. Different numbers of observations result from the fact that not everybody uses a car.

Regarding the third working hypothesis and analogous to the SOEP waves of 1998 and 2003, per capita CO_2 emissions stemming from individual transportation were calculated from fuel type, fuel combustion per 100 km, and annual mileage for each reported car. In the case of carbon emissions resulting from building operations, households were asked to report all energy carriers used in the household as well as respective monthly or annual energy expenditures. Energy expenditures were translated into absolute amounts which, in turn, were converted to CO_2 emissions. With regard to consumption-triggered CO_2 emissions, only incomes (i.e. household incomes minus energy expenditures in the realms of housing and mobility) are reported here. As discussed in the previous chapter, the procedure had to be formulated in this way because no reliable tool has been developed to estimate how incomes translate into carbon emissions.

Here again, and given the central research question inspired by fractionalization, a mainly descriptive analytical approach appears to suffice. Thereby, and analogous to subchapter 5.2, Cronbach's alpha (α) as well as bivariate correlations (Pearson's correlation coefficient, r) are used in order to check the reliability of the selected variables measuring dwelling technologies or cultural preferences, and to examine whether the theoretically assumed correlations among these variables within urban paths of adaptation actually exist.

5.3.3 Empirical evidence for rural and urban paths of adaptation within cities

The central findings regarding the correspondences among dwelling technologies, cultural preferences, modes of social reproduction, and related CO_2 emissions in Munich and Bolzano are summarized in Table 5.5. In order to account for city-specific peculiarities, however, the following discussion also refers to Tables A.2 and A.3, where key findings for Munich and Bolzano are shown, respectively. In general, and when compared with the differences between rural and urban paths of adaptation as discussed in the previous chapter, the findings presented here are more ambiguous. This is also indicated by rather weak Cronbach's alpha values which – here again – are used as indicators evaluating the validity of measuring "dwelling technologies" ($\alpha = 0.35$), "cultural preferences" ($\alpha = 0.59$), and "modes of social reproduction" ($\alpha = 0.52$) with the help of the reported variables. It is likely that low Cronbach's alpha values occur because observable variation within agglomerations is not as pronounced as differences between rural and urban paths of adaptation. In addition, the variation which can be observed on the household level is always higher than that of aggregated data, and therefore Cronbach's alpha values should not be compared. Nevertheless, some parallel effects between rural and urban paths of adaptation and dispersion vs. connected dwelling technologies within urban agglomerations seem to appear. As a result of missing generalizability and weak Cronbach's alpha values, the following discussion has some reservations and – for that reason – is kept comparatively short.

The first working hypothesis postulates that physically connected dwelling technologies correspond with social-democratic value orientations and a comparatively broader range of family and life plans, whereas unconnected dwelling technologies should be accompanied by rather traditional-conservative value orientations and a narrower and more homogeneous scope of life plans. To begin with, and only regarding dwelling technologies, similar effects to the case of rural and urban paths of adaptation can be observed. Despite some city-specific peculiarities, the general trend is that the relative frequencies of grid-independent energy carriers such as oil decline with increasing building size, as do living space per capita and household size. Thereby, the widest differences appear between the first three and the last two building types. Only the number of cars and number of households using gas as the main energy carrier appear to be independent of building type within urban agglomerations.

Looking at corresponding cultural preferences, Table 5.5 shows that detached housing clearly corresponds with a high share of property owners, political preferences represented by center-right parties, a high proportion of marriages, and a strong willingness to support relatives in case of illness or the infirmities of old age. To give some examples from the bivariate correlation matrix, the observable correspondences between building type and real-estate ownership (indicating possessive individualism) are always significant (with significance levels ranging from $p < 0.001$ to $p < 0.05$). Political preferences represented by center-right parties correlate weakly positively with detached housing ($r = 0.07$; $p < 0.01$) and oil as the main energy carrier ($r = 0.20$; $p < 0.01$). Apartment buildings, in contrast, clearly correspond with a comparatively higher proportion of tenants, left-wing political preferences, a higher share of singles and unmarried couples, and a much weaker willingness to support care-dependent relatives. However, bivariate correlations are weak or insignificant here. Nevertheless, the figures presented in Table 5.5 (as well as in Tables A.2 and A.3) – especially those concerning real-estate ownership, political preferences, and marital status – indicate that detached forms of dwelling are actually accompanied by comparatively more traditional-conservative cultural preferences. More compact forms of dwelling coincide with more heterogeneous political preferences and a broader range of private life plans. Based on these observations, and considering the constraints discussed above, the first working hypothesis could be seen as tentatively approved.

Focussing on different modes of social reproduction, the second working hypothesis claims that forms of detached housing (i.e. the first three building types) coincide with comparably lower degrees of labor division (and lower incomes), more "do it yourself work", and a larger proportion of children, whereas compact forms of dwelling should correspond with higher degrees of labor division, higher incomes, and fewer children. Thus, occupiers of (semi-)detached or row houses should feature, for example, lower educational levels, lower incomes, and more children than those living in apartment buildings.

In a first attempt, no systematic trend between building type, cultural preferences, and modes of social reproduction could be identified. Probably, and as indicated by the average age as well as by the large proportion of retired men and women, this is partly due to the fact that the samples are strongly age-biased (especially the Munich sample). On this basis, it does not make sense to search for systematic patterns between different forms of intra-familial labor division and related numbers of children on the one side and building type, dwelling technologies, or cultural preferences on the other. For that reason, the following considerations are restricted to gainfully employed persons.

The data reveal some of the expected systematic patterns (even though not as pronounced as in subchapter 5.2). Regarding educational levels, the working hypothesis cannot be affirmed. Irrespective of building type and gender, educational levels are high. This can probably be explained by self-selectivity, meaning that well-educated individuals are more interested in ecological issues and – for that reason – showed a higher willingness to participate in the survey. Apart from this, educational levels tend to be higher in urban areas.

Concerning employment rates, similar patterns as identified in the comparison of rural and urban paths of adaptation (cf. subchapter 5.2) can be observed. Regardless of building type, almost all men work full-time. In contrast, the proportion of women working full-time tends to increase with building size whereas the number of women working part-time decreases with increasing building size. This observation tentatively underlines the findings described above: detached forms of dwelling are accompanied by comparatively more traditional-conservative cultural preferences, which also includes the fact that women do not work or, if at all, work only part-time; more compact forms of dwelling coincide with comparatively more heterogeneous political preferences and a broader range of private lifestyles, including a comparatively large proportion of women working full-time. Last but not least, and with regard to household income and the number of children, the working hypothesis can only be partially confirmed. Detached forms of housing actually correspond with comparatively lower incomes and more children in Bolzano, but no clear-cut trend can be identified for Munich.

Taken together, it is questionable whether the empirical findings regarding the second working hypothesis can be trusted and generalized – especially those concerning household incomes. If at all, it is assumed that these findings primarily apply to a well-educated middle class (which apparently dominates the samples used here). Clearly more research with more appropriate data has to be done to definitively assess the correspondences among different building types and modes of social reproduction within urban agglomerations.

Finally, the third working hypothesis postulates that per capita CO_2 emissions stemming from individual transportation and building operations should decline with increasing degrees of physical connectedness. As illustrated in Table 5.5, this thesis is supported by the present data. Although the number of cars does not significantly vary with building type, respondents practicing detached forms of dwelling apparently emit more carbon from individual transportation than those living in apartment buildings. This can probably be explained by the fact that physically compact forms of dwelling correspond with infrastructural context-variables such as public transportation or pedestrian-friendly local supply structures (not reported here), whereas detached forms of housing go hand in hand with infrastructural properties more or less requiring the use of individual transportation. In the case of CO_2 emissions resulting from building operations, the observable differences between detached and physically compact building types can be explained by the fact that more compact buildings are marked by superior SA:V ratios and less living space per capita. Analogous to the findings of subchapters 5.1 and 5.2, the effects of physical and social concentration work in opposite directions within urban agglomerations.

With regard to consumption-triggered CO_2 emissions, it was assumed that household incomes would increase with rising degrees of physical compactness. At least at a first glance, this is actually the case. As detached forms of dwelling are accompanied by comparatively lower per capita household incomes and higher energy expenditures (due to more living space per capita and unfavorable SA:V ratios), those households residing in apartment buildings apparently have higher incomes at their disposal (per capita). However, it has to be considered that rental charges were not surveyed, but should be subtracted from incomes in the case of tenants – otherwise, income-triggered CO_2 emissions of tenants are overvalued. Thus, actual differences in incomes should be more balanced than suggested by Table 5.5. In view of these uncertainties, more research has to be done (using more appropriate data) before the correspondence between building type and consumption-triggered CO_2 emissions in urban agglomerations can be explained in a more satisfactory way.

5.3.4 Concluding remarks

Against the theoretical background of fractal geometry (Mandelbrot 1967, Mandelbrot 1987, Batty 2011, 2013, Bettencourt 2013), the central goal of the present chapter was to determine whether similar technological, cultural, and economic differences characterizing rural and urban paths of adaptation could be observed within urban settlement structures. All in all, the analyses presented above confirm this assumption. In particular, those correspondences mediated by the universally valid laws of allometry and economies of scale can also be observed within urban paths of adaptation. The correspondences among building type, dwelling technologies, and cultural preferences seem to be fairly clear, too. Comparable to rural paths of adaptation as described in subchapter 5.2, detached forms of (intra-urban) dwelling go hand in hand with a rather homogenous set of traditional-conservative and family-oriented cultural preferences and related modes of social reproduction. More compact forms of dwelling, in contrast, coincide with a rather heterogeneous set of social-democratic and more open-minded cultural preferences and a broader scope of related family and life plans.

The assumed correspondences among dwelling technologies, cultural preferences, and related modes of social reproduction, by contrast, were confirmed only tentatively. However, this does not necessarily mean that no correspondence exists as predicted by the second working hypothesis. To give an example, this is also indicated by the (theoretically expected) fact that a comparatively low proportion of female full-time employment coincides with detached forms of housing and related conservative-traditional cultural preferences. Thus, it can be assumed that more appropriate data (for example, census-type data as used in subchapter 5.2) would yield the theoretically expected results.

From a theoretical and methodical perspective, the analyses once again reveal that the approach of eco-cultural adaptation is operationalizable and in most cases yields the expected results. Moreover, and entirely in line with the idea of fractionalization, this chapter also demonstrates that the theoretical framework developed in Chapters 2 and 3 is sufficiently universal to be applied to different spatial realms without compromising predictive power.

5.4 The concept of eco-cultural adaptation: a robust analytical tool

The previous sections should be seen as the very first empirical attempts at utilizing the concept of eco-cultural coevolution as developed in Chapters 2 and 3. Taking as an example different dwelling technologies in rural and urban paths of adaptation, the foregoing aimed to demonstrate that some (theoretically predictable) matches among technologies, cultural preferences, and related modes of social reproduction are observable – more frequently than by random chance – in rural as opposed to urban settlement structures. Accordingly, not only predictive accuracy but also crucial questions regarding the operationalization of key components of the eco-cultural fabric (at different spatial scales) were first put to the test. So, what overall theoretical and methodical conclusions can be drawn? What can be achieved by the analyses presented? What issues should be put on the future research agenda?

From a theoretical and practical-methodical perspective, the concept of eco-cultural adaptation is proven as a robust analytical tool. Not only does it allow the deduction of working hypotheses, but also their quantitative-empirical verification. The previous subchapters revealed that operationalizing fundamental components of eco-cultural habitats is a feasible endeavor – even on different analytical scales – and (mostly) yields the theoretically expected results. Thus, it could be argued that the concept of eco-cultural adaptation allows the formulation of generalizable statements about the relationship between physical nature and society – a quality claimed for only a few theoretical concepts problematizing the mutual interaction between physical environs and society.

Despite these positive overall impressions, the previous chapters have also disclosed some open questions and related fields of further research. To begin with, the spatial distribution of CO_2 emissions in Bavarian municipalities suggests that a one-dimensional theoretical approach accounting only for differences in physical compactness apparently does not fully suffice to explain observable spatial variations. Here, integrating a second dimension accounting for market proximity and – in doing so – accounting for spatial differences in labor division, incomes, and related levels of income-triggered resource consumption represents a crucial field of further research. As discussed above, this endeavor raises serious questions regarding the translation of monetary values into ecological impacts. As elaborated by Hertwich (2005), this means not only that future approaches have to account for spatial differences in consumption patterns and related baskets of commodities (for example, costly haircuts or carbon-intensive air travel), but consideration must be given to the role of imports and related ecological impacts of foreign production conditions. Given these complex requirements, the success of future research endeavors largely depends not only on appropriate data, but also on the methodological question of how deep the research should go without getting lost in the ramifications of material flow analysis (ibid.: 4681 ff.).

Critics could object that only Western dwelling technologies and related eco-cultural fabrics were scrutinized. To determine whether the concept of eco-cultural adaptation can be described as an ahistorical and universally valid theoretic frame of reference, the examination of different dwelling technologies and related eco-cultural paths of adaptation in various ecological and socio-economic environments should be put on the future research agenda, for example, in emerging economies or agrarian societies (Bevan 2004, Wood 2004, Gough 2007).

Moreover, and more fundamentally, criticism could be levelled at the fact that the analyses presented here predominantly use cross-sectional data (meaning that only one point in time was examined), whereas the approach of eco-cultural adaptation is claimed to be able to explain temporal changes. In other words, previous analyses do not account for the fact that changing physical environments, technologies, institutions, or cultural preferences – in the course of time – will provoke mutations in all other components of the respective eco-cultural fabric. Admittedly, this is a crucial objection. To do justice to the temporal aspect (among others), Chapter 6 will exclusively deal with the phenomenon of emerging eco-cultural habitats in a reconstructive-narrative way. As demonstrated by Sieferle et al. (2006), however, a quantitative analysis of habitat emergence and transformation is also feasible – but only on the condition that appropriate longitudinal data are available. Keeping with the example of dwelling technologies and related cultural preferences and modes of social reproduction, longitudinal data covering the spatial distribution of infrastructural and cultural properties, as well as data concerning labor division and related modes of social reproduction since the early nineteenth century (for example, with sampling intervals every 25 years), would offer an appropriate basis for an analysis of the dynamics of habitat-emergence and transformation. Here, it would be interesting to explore how technological changes (keyword "industrial revolution") – over the course of time – correspond with infrastructural changes (for example, increasing degrees of urbanization) and related economic and socio-demographic shifts and upheavals. However, collecting such data represents a research project in its own right, not to mention the analyses that are required. Therefore, available cross-sectional data were used here to identify contemporary constellations among dwelling technologies, cultural preferences, and related modes of social reproduction. In view of the promising overall findings of this chapter, a research program collecting longitudinal data as sketched out above and aimed at verifying or disproving the approach of eco-cultural adaptation should definitely be put on the future research agenda.

However – to come back to the present – the issues of habitat-emergence and transformation are addressed in an historical and reconstructive-narrative way in the next chapter. Particularly, and taking as examples the Netherlands and the United States (especially New Orleans, Louisiana), the following questions will be addressed: how do eco-cultural paths of adaptation (and related matches among physical natures, technologies, institutions, and cultural preferences) emerge over the course of time? How are physical natures, technologies, institutions, and cultural preferences interlinked? Is it true that physical environs and social relationships are constituted in similar ways? And, last but not least, how do the liberal and coordinated cultural biases affect the adaptational capacities of US and Dutch eco-cultural habitats today?

Notes

1 Subchapter 5.1 was originally published in *Environmental Policy and Governance* in a slightly different version (Schubert et al. 2013).
2 In Germany, rural and urban settlement structures tap the "subterranean forest" (Sieferle 2010). In contrast to physically compact towns, however, the use of fossil fuels had to be integrated into historically grown dispersed settlement structures in rural areas. Therefore, grid-bound energy carriers are more common in urban areas, whereas rural areas still depend on unconnected energy carriers such as oil, coal, or firewood.
3 Residents of New York City on average emit only 10.5 t CO_2/capita per year, whereas the nationwide average is 23.6 t CO_2/capita (Hoornweg et al. 2011).
4 Different forms of scaling (for example, the relationships among body mass, body size, anatomy, and metabolic rate) have been explored in evolutionary biology and statistics in order to determine universally valid rules and power law functions to express these relationships in the form of scaling exponents ($y = C a^{-\beta}$) (Schmidt-Nielsen 1984, Vogel 1988, West et al. 1997). C is the intercept of the function and is usually of minor interest, since it is simply a scaling parameter multiplied by the core power law function a^{β}. Thus, β is the main factor influencing the functional relationship between a and y. Values above 1 imply that y increases at a faster rate than a (positive allometry). Values below 1 imply that y increases at a slower rate than a (negative allometry).

5   Among other things, this goes back to the properties of medieval towns which, in terms of size, had to be rather small in order to allow for shorter city walls (which were cheaper to raise and easier to defend than long walls).

6   Here again, it could be argued that these structural properties are remnants of an ancient metabolic regime based on solar energy stored in biomass.

7   Although the results presented here look quite convincing, it is important to have in mind that it is rather difficult to survey costs for heating and mobility. Many people do not keep their heating and electricity bills or are unable to read them properly (Frondel 2005). Given these problems, the reliability of these results should be validated with other datasets in the future (cf. subchapters 5.2 and 5.3).

8   In contrast to direct emission factors, total emission factors also include all energetic inputs that accumulate during the manufacturing process (for example, drilling, transport, and refining).

9   As these findings are robust over the years, only the results that apply to 2003 (cf. Table 5.2) are reported. See Appendix to find out more about the results that apply to 1998 (Table A.1).

10   Unfortunately, some of the most relevant variables (for example, information about the heating system, mileage per year, or energy costs) are only available for two points in time, namely 1998 and 2003. That is why fixed effects models can only be estimated for utilities, but not for carbon emissions.

11   The CO_2-neutral use of firewood was not surveyed. Nevertheless, it can be assumed that it is more frequently used in rural than in urban areas.

12   Note that emissions resulting from public transportation are not accounted for here. Because public transport is more commonly used in urban areas, the figures presented here underestimate the per capita carbon emissions of city dwellers.

13   The dataset was compiled in the course of the triannual research project "Klima Regional – soziale Transformationsprozesse für Klimaschutz und Klimaanpassung" funded by the German Ministry of Education and Research (www.klima-regional. de). The data were sourced from the following institutions (without any claim to completeness): Bayerisches Landesamt für Statistik, Bayerisches Landesamt für Vermessung und Geoinformation, Bundesagentur für Arbeit, Bundesinstitut für Bau-, Stadt- und Raumforschung, Leibniz-Institut für ökologische Raumentwicklung, Kraftfahrt-Bundesamt, Deutscher Wetterdienst.

14   Regarding educational qualifications and employment rates, only those people subject to social insurance contributions were captured here – self-employed people or mini-jobbers were not included.

15   The nationwide value of 5.58 t CO_2/capita had to be statistically weighted to account for the fact that Bavarian incomes, on average, are approximately 6.6% above the national average.

16   In this context, however, it is important to remember that the intercorrelation among items increases with increasing numbers of items. Thus, high Cronbach's alpha values indicating strong internal consistency in the case of large numbers should always be critically reflected.

17   www.klima-regional.de

Bibliography

Allison, P. D. (2009). *Fixed effects regression models.* Quantitative applications in the social sciences: Vol. 160. Thousand Oaks: Sage.

Angrist, J. D., & Pischke, J.-S. (2009). *Mostly harmless econometrics: An empiricist's companion.* Princeton: Princeton Univ. Press.

Batty, M. (2011). Building a science of cities. *Current Research on Cities*, 29, Supplement 1, 9–16.

——— (2013). A theory of city size. *Science*, 340(6139), 1418–1419.

Bettencourt, L. M., Lobo, J., Helbing, D., Kühnert, C., & West, B. (2007). Growth, innovation, scaling, and the pace of life in cities. *Proceedings of the National Academy of Sciences*, 104(17), 7301–7306.

Bevan, P. (2004). Conceptualising in/security regimes. In I. Gough (Ed.), *Insecurity and welfare regimes in Asia, Africa, and Latin America. Social policy in development contexts* (1st ed., pp. 88–118). Cambridge: Cambridge Univ. Press.

Binswanger, M. (2001). Technological progress and sustainable development: what about the rebound effect? *Ecological Economics*, 36(1), 119–132.

Cole, M. A., & Neumayer, E. (2004). Examining the impact of demographic factors on air pollution. *Population and Environment*, 26(1), 5–21.

Diekmann, A. (2008). *Empirische Sozialforschung: Grundlagen, Methoden, Anwendungen*. Reinbek: Rowohlt.

DIW – Deutsches Institut für Wirtschaftsforschung (2011). www.diw.de/en/diw_02.c.221178. en/about_soep.html, (Sep. 2, 2011).

Ellickson, R. (1993). Property in Land. *Faculty Scholarship Series*, Paper 411. http:// digitalcommons.law.yale.edu/fss_papers/411

Federal Ministry of Economics and Technology 2011. *Entwicklung von Energiepreisen und Preisindizes*, Energiedaten Tabelle 26a, http://bmwi.de/BMWi/Navigation/Energie/ Statistik-und-Prognosen/energiedaten.html (Nov. 8, 2011).

Fischer-Kowalski, M. (Ed.) (1997). *Gesellschaftlicher Stoffwechsel und Kolonisierung von Natur. Ein Versuch in sozialer Ökologie*. Amsterdam: Fakultas.

Fritsche, U. R. (2007). *Energiebezogene Gesamtemissionen für Treibhausgase aus fossilen Energieträgern unter Einbeziehung der Bereitstellungsvorketten, Kurzbericht im Auftrag des Bundesverbands der deutschen Gas- und Wasserwirtschaft e.V. (BGW)*. Öko-Institut Darmstadt.

Fritsche, U.R., & Rausch, L. (2007). *Bestimmung spezifischer Treibhausgas-Emissionsfaktoren für Fernwärme*. Öko-Institut Darmstadt.

Frondel, M. (2005). *Erhebung des Energieverbrauchs der privaten Haushalte für das Jahr 2005*. RWI Essen.

Galilei, G. (1730 [1638]). *Mathematical discourses concerning two new sciences relating to mechanicks and local motion, in four dialogues. ... By Galileo Galilei, ... With an appendix concerning the center of gravity of solid bodies. Done into English from the Italian, by Tho. Weston, ... and now publish'd by John Weston;*. London: printed for J. Hooke.

Gemis 2011. Global Emission Model for Integrated Systems (GEMIS) Version 4.7. www. oeko.de/service/gemis/en/index.htm (Nov. 4, 2011).

Glaeser, E. L. (2011). *Triumph of the city: How our greatest invention makes us richer, smarter, greener, healthier, and happier*. New York: Penguin Press.

Glaeser, E. L., & Kahn, M. E. (2010). The greenness of cities: Carbon dioxide emissions and urban development. *Journal of Urban Economics*, 67(3), 404–418.

Gough, I. (Ed.) (2007). *Wellbeing in developing countries: From theory to research* (1st ed.). Cambridge: Cambridge Univ. Press.

Greene, W. H. (2003). *Econometric analysis* (5th ed.). Upper Saddle River, NJ: Prentice Hall.

Gujarati, D. N. (2003). *Basic Econometrics*. Boston: McGrawHill.

Haldane, J. B., & Maynard Smith, J. (1985). *On being the right size and other essays*. Oxford: Oxford Univ. Press.

Häußermann, H., & Siebel, W. (1996). *Soziologie des Wohnens*. Weinheim: Juventa.

Henrich, J., Boys, R., Bowles, S., Camerer, C., Fehr, E., & Gintis, H. (Ed.) (2004). *Foundations of human sociality: Economic experiments and ethnographic evidence from fifteen small-scale societies*. Oxford: Oxford Univ. Press.

Hertwich, E. G. (2005). Life cycle approaches to sustainable consumption: A critical review. *Environmental Science & Technology*, 39(13), 4673–4684.

Holden, E., & Norland, I. (2005). Three challenges for the compact city as a sustainable urban form: Household consumption of energy and transport in eight residential areas in the greater Oslo Region. *Urban Studies*, 42(12), 2145–2166.

Hoornweg, D., Sugar, L., & Trejos Gomez, C. L. (2011). Cities and greenhouse gas emissions: moving forward. *Environment and Urbanization*, 23(1), 207–227.

Jackson, T. (2009). *Prosperity without growth: Economics for a finite planet*. London: Earthscan.

Lähteenoja, S,. Lettenmeier, M., & Kotakorpi, E. 2008. *The ecological rucksack of households – huge differences, huge potential for reduction?* Proceedings: Sustainable Consumption and Production: Framework for action, 10–11 March 2008, Brussels, Belgium. Conference of the Sustainable Consumption Research Exchange (SCORE!) Network, supported by the EU's 6th Framework Programme. 1–21.

Mandelbrot, B. (1967). How long is the coast of Britain? Statistical self-similarity and fractional dimension. *Science*, 156(3775), 636–638.

(1987). *Die fraktale Geometrie der Natur*. Basel: Birkhäuser.

Mcpherson, C. (1962): *The political theory of possessive individualism: From Hobbes to Locke*. Oxford: Clarendon.

North, D. C. (1990). *Institutions, institutional change, and economic performance*. Cambridge: Cambridge Univ. Press.

Otte, G., & Baur, N. (2008). Urbanism as a way of life? Räumliche Variationen der Lebensführung in Deutschland. *Zeitschrift für Soziologie*, 37(2), 93–116.

Owen, D. (2009). *Green metropolis: Why living smaller, living closer, and driving less are keys to sustainability*. New York: Riverhead Books.

Schmidt-Nielsen, K. (1984). *Scaling, why is animal size so important?* Cambridge: Cambridge Univ. Press.

Schubert, J., Wolbring, T., & Gill, B. (2013). Settlement structures and carbon emissions in Germany: The effects of social and physical concentration on carbon emissions in rural and urban residential areas. *Environmental Policy and Governance*, 23(1), 13–29.

Sieferle, R. P. (2010). *The subterranean forest*. Cambridge: White Horse Press.

(2011). Cultural evolution and social metabolism. *Geografiska Annaler: Series B, Human Geography*, 93(4), 315–324.

et al. (Ed.) (2006). *Das Ende der Fläche: Zum gesellschaftlichen Stoffwechsel der Industrialisierung*. Köln: Böhlau.

UBA – Umweltbundesamt (2013). http://uba.klimaktiv-co2-rechner.de/de_DE/page/ (Aug., 05. 2013).

VandeWeghe, J. R., & Kennedy, C. (2007). A spatial analysis of residential greenhouse gas emissions in the Toronto census metropolitan area. *Journal of Industrial Ecology*, 11(2), 133–144.

Vogel, S. (1988). *Life's devices: The physical world of animals and plants*. Princeton: Princeton Univ. Press.

Weber, C., & Perrels, A. (2000). Modelling lifestyle effects on energy demand and related emissions. *Energy Policy*, 28(8), 549–566.

West, G. B., Brown, J. H., & Enquist, B. J. (1997). A general model for the origin of allometric scaling laws in biology. *Science*, 276(5309), 122–126.

Wood, G. (2004). Informal security regimes: the strength of relationships. In I. Gough (Ed.), *Insecurity and welfare regimes in Asia, Africa, and Latin America. Social policy in development contexts* (1st ed., pp. 49–87). Cambridge: Cambridge Univ. Press.

6 Social welfare and flood protection in river deltas

Trying to understand why ecological and social challenges such as flooding or social inequality are dealt with differently in the Rhine–Meuse–Scheldt and Mississippi deltas, this chapter first makes an attempt to intertwine the concept of eco-cultural adaptation with the typology of welfare and production regimes put forward by Gøsta Esping-Andersen (1990) and Hall and Soskice (2001), and corresponding types of environmental regulation as observed by, for example, James Meadowcroft (2005) or Andreas Duit (2008). The central idea of this chapter is that different approaches to social and ecological challenges can be explained by distinct forms of path-specific institutional specialization and cultural persistency, which, in turn, should be reflected in distinct matches between physical nature, technologies, institutions, and cultural preferences. Second, paradigmatic examples illustrating how these different approaches might have historically emerged and affected past, and still affect present, approaches to flood protection and social welfare are presented and theoretically reflected.

In view of the fact that both deltas share many crucial geomorphological and socio-economic similarities – especially with regard to their physical surroundings as well as available know-how, technologies, or financial resources – different approaches to physical and social challenges are all the more astonishing. Both habitats are located in deltaic wetlands characterized by a low elevation above sea level (or even below) and land subsidence. Therefore, they are highly vulnerable to floods either from the sea (the North Sea and Gulf of Mexico, respectively), from inundation (the Rhine and Mississippi rivers, respectively), or strong precipitation (Colten 2000, 2005, van de Ven 2004, Hudson et al. 2008, Syvitski et al. 2009). Further, it can be argued that both habitats have comparable know-how and financial resources at their disposal and use similar hydraulic works – levees, ditches, sluices, water pumps, and the like – to create and protect habitable land.

Considering these similarities, the devastation of New Orleans by Hurricane Katrina in 2005 revealed that the city – once again – was badly prepared to cope with extreme weather and its humanitarian aftermath (Davis 2000, Hartman/ Squires 2006, Bates/Swan 2007, Brunsma et al. 2010). Approximately 1,800 people – mostly black people of low socio-economic status – were killed and by 2007, more than one third of those forced to leave New Orleans (approximately

175,000 people) had not returned to the city (Fussell et al. 2010). In contrast, and except for the North Sea flood of 1953, which resulted in 1,836 fatalities, the Netherlands have mostly been spared severe flooding for more than 50 years (Lamb 1991, van de Ven 2004). So, how can these differences be explained? Usually, historically singular aspects such as technical and human failure or extraordinary meteorological conditions resulting in 100-year floods are used to explain the devastating effects of extreme weather events such as hurricanes. However, considering the astonishing geomorphological and technological similarities between New Orleans and the Netherlands and taking into account the long-lasting (but arguably inadequate) attempts to establishe sufficient hydraulic works along the Mississippi and in New Orleans (Colten 2000, 2005, 2009), it looks as if explanations taking recourse to historically unique and fatal coincidences of unfortunate circumstances miss the central point here: fundamental structural mismatches between flood protection, hydraulic works, and social welfare on the one side and essential institutional and cultural necessities regarding their provision and maintenance on the other.

It was argued that eco-cultural habitats can be defined as specific constellations between human society and its physical surroundings. From an ideal-typical perspective, these constellations are mediated via fit matches between technologies, institutions, and cultural preferences. Against this theoretical background and due to the fact that the Rhine–Meuse–Scheldt and the Mississippi deltas are characterized by comparable physical environments and comparable technological solutions to flooding, it should be possible to explain different approaches to ecological and social challenges arising in the realms of flood protection with the help of distinct institutional structures and related cultural preferences. In consequence, and in order to formulate a tentative working hypothesis, the ensuing discussion will revolve around the following assumption: the devastation of New Orleans by Hurricane Katrina and its humanitarian aftermath is nothing exceptional, but simply offers another occasion to study the structural mismatches between a liberally biased mode of life on the one side and the institutional and cultural necessities associated with the provision and maintenance of public goods on the other (Schreuder 2001). To put it more generally, it is assumed that non-excludable goods such as flood protection (but also clean air, public security, public health, or public education) suffer in liberally biased institutional and cultural settings, but flower in eco-cultural habitats characterized by strong preferences for cooperative behavior and hierarchic coordination.

Given this central research idea, the ongoing discussion will proceed as follows: in a first step, and using the relevant literature in relation to different welfare and production regimes as well as corresponding forms of environmental regulation, different regulatory approaches coordinating the utilization of physical and social relationships are presented and paradigmatically compared. The aim is to find some empirical evidence corroborating the central theoretical assertion according to which the relationships among social actors, as well as their interference with physical nature, proceed in accordance with one and the same set of path-specific institutional structures and cultural preferences. Accordingly, and true to the motto "If you only have a hammer, every problem looks like a nail" (Maslow 1966), ecological and social challenges should be similarly approached within one and the same habitat of eco-cultural adaptation.

In a second step, and based on these conceptual considerations, the analytical perspective shifts from different institutional environments and corresponding cultural preferences to the concrete and tangible level of everyday life. Thereby, the questions of how path-specific forms of institutional and cultural specialization as described above historically gained shape and still affect daily routines and practices occupy center stage. As elaborated in Chapters 2 and 3 and comparable with Giddens' theory of structuration (Giddens 1984), it is argued that institutional structures and related cultural preferences provide social actors with ready-made solutions to given tasks and challenges in everyday life and – for that reason – leave long-lasting imprints on their capacities of coping with social or ecological tasks and challenges (Watts/Bohle 1993, Bohle et al. 1994, Adger/Kelly 1999).

For illustration, past and present approaches to flood protection and social welfare in the Rhine–Meuse–Scheldt and Mississippi deltas are exemplarily dealt with. To be more precise, the emergence of path-specific institutional and cultural specialization is illustrated against the background of flood protection in the Netherlands. The ambivalent effects of institutional and cultural specialization on habitat-specific adaptation capacities are illustrated against the example of New Orleans and its environs.

6.1 Similar approaches for regulating nature and society

The present section aims to identify different regulatory approaches coordinating the utilization of physical nature and society in the Rhine–Meuse–Scheldt and Mississippi deltas and could thus be described as "institutional inventory". For this purpose, the concept of eco-cultural adaptation is intertwined with the typology of welfare and production regimes as put forward by Esping-Andersen (1990) and Hall and Soskice (2001), as well as with related regulatory modes organizing humans' interference with physical nature as described, for example, by Scruggs (1999), Meadowcroft (2005), or Duit (2008). Thereby, it is assumed that ecological and social tasks and challenges are coordinated by one and the same set of institutional structures and corresponding cultural preferences (i.e. elective affinities). Before going into detail, however, central arguments providing theoretical reasons for similar regulatory approaches to social and physical nature are briefly recapitulated.

Depending on the prevailing costs and benefits of labor division, it was argued that – in the course of time – cultural preferences emerge favoring either solitary action or cooperative behavior (Henrich et al. 2004, Bednar/Page 2007, Henrich/ Henrich 2007). Once established, and in order to reduce cognitive burdens of situational decision making and related learning costs, these cultural preferences and behavioral strategies are applied to different tasks or challenges. As time goes by, path-specific cultural preferences thus become more and more coherent and homogenous, which, in turn, results in an overall increase in societal efficiency by fostering mutual understanding and expectability. In the long run, this leads to a constellation in which social interaction is characterized by "intra- and inter-agent behavioral consistency" and where "rational agents choose (for rational reasons) to act culturally" (Bednar/Page 2007: 66).

These considerations – usually applied only to the relationships and interactions among social actors – can be easily transferred to humans' interference with nature: first of all, it was argued that transferring one and the same set of cultural preferences (and related behavioral strategies) to the coordination of human–nature relationships should be cognitively more efficient than situationally switching to and fro between a multitude of different object-specific heuristics. Second, and in the case of high benefits to long-term cooperation and reputation building, it is not only wise to treat one's fellow men well, but also the physical environment in which they live and on which they are dependent. By implication, this means that low paybacks to reputation building and cooperation result in less frequent interaction, which, in turn, should become observable in comparably less cautious environmental behavior. Last but not least, and due to path-specific institutional specialization, the assumption should be allowed that social and ecological challenges are approached in the same way. To give an example, it can be assumed that eco-cultural habitats featuring an institutional environment that has emerged in order to support solitary action and coordination by means of (market) competition will utilize market incentives in order to shape humans' interference with nature.

The discussion in subchapter 3.1 came to the following conclusion: social groups populating eco-cultural habitats primarily specialized in cooperative action are not supposed to make any marked difference between social cohesion, non-kin solidarity, and environmental sustainability and – for that reason – cautiously treat their physical environment like fellow men. In contrast, individuals inhabiting eco-cultural habitats mainly specialized in establishing institutional circumstances allowing for solitary action and competitive behavior are expected to instrumentalize both fellow men and physical nature for the sake of short-term maximization of individual utility.

Given this brief theoretical repetition, what does this have to do with welfare and production regimes and corresponding modes of interfering with physical nature? To begin with, and borrowing from Karl Marx (1932a, 1932b) and Sieferle et al. (2006), it can be argued that human society regulates and maintains its material and energetic exchange relations with physical nature – that is, its social metabolism – via labor and production processes in the broadest sense, for example, by extracting raw materials and processing them into nourishment, shelter, and other necessities (Sieferle 2011). In other words, labor and production processes are fundamental elements of human existence through which physical nature and manpower are appropriated and utilized according to human needs and wants. Thus, having a closer look at the specifics of predominant labor and production processes should emerge as a worthwhile endeavor when trying to learn more about how physical nature and social society are approached in distinct eco-cultural habitats of adaptation.

Searching for an economic synonym capable of catching the meaning of social metabolism, "economy" comes to mind. "In its most fundamental sense, an *economy* is *a collection of labor processes*" (Bowles et al. 2005: 97). Today, the predominant economic order – that is, the predominant form of organizing labor processes – is capitalism, which is defined as "an economic system in which employers, using privately owned capital goods, hire wage labor to produce commodities for the purpose of making a profit" (ibid.: 4). Although this definition might suggest that capitalism represents itself as a homogenous system, different varieties of capitalism can be empirically observed (Hall/Soskice 2001).[1] These differences are deeply rooted in distinct institutional structures and corresponding regulatory regimes into which labor and production processes are embedded and which crucially affect a firm's performance.[2]

From an ideal-typical perspective, liberal and coordinated market economies are distinguished: "In *liberal market economies*, firms coordinate their activities primarily via hierarchies and competitive market arrangements" (ibid.: 8), whereas firms in coordinated market economies "more heavily depend on non-market relationships to coordinate their endeavors with other actors and to construct their core competencies" (ibid.: 8). In this context, it is important to note that the term "hierarchy" refers to hierarchies within a firm and should not be confused with centralized forms of hierarchic regulation as described, for example, by Wittfogel (1957).

However, labor and production processes depend not only on institutional arrangements guiding their coordination in and between firms, but also on institutional structures regulating the valorization of manpower and physical nature in a much broader sense – for example, with regard to issues such as education, sick pay, and dismissal protection or emission limits, the establishment of preserve areas, and the storage and disposal of waste. Thus, the following discussion will focus on elective affinities between liberal and coordinated production regimes on the one side and corresponding institutional arrangements organizing the utilization of manpower and physical nature on the other. Thereby, empirical evidence corroborating the theoretical idea according to which a limited and homogenous set of cultural and coordinational principles permeates any realm of social action will be revealed.

6.1.1 Welfare regimes and the valorization of human labor

To begin with, and as observed by Martin Schröder (2009), liberal and coordinated market economies correspond with suitable welfare arrangements regulating the utilization of human labor. Trying to establish theoretical and empirical links between liberal and coordinated production regimes as put forward by Hall and Soskice (2001) on the one side and the typology of welfare states as introduced by Esping-Andersen (1990), Schröder (2009) makes the crucial point that these typologies are closely related, both theoretically and empirically:

Whereas VOC [varieties of capitalism, author's note] wants to understand how firms deal with institutional environments that vary between production systems of different countries, Esping-Andersen analyses modes according to which welfare is distributed based on rights and duties of individuals *vis-à-vis* the state. Importantly, though, both typologies arrive at very similar country groupings, since "virtually all liberal market economies are accompanied by 'liberal' welfare states" (Hall and Soskice 2001: 50) and all coordinated market economies are accompanied by either a social democratic or a conservative welfare arrangement.

Schröder (2009): 21

The main criterion to distinguish between liberal and coordinated welfare regimes is the degree of decommodification a nation state provides for its citizens. This measures the extent to which social security depends on gainful employment (Esping-Andersen 1990).[3] Liberal welfare regimes such as the United States are marked by a comparably low degree of decommodification, meaning that only a minimum standard of public services and social welfare is provided by the state. Anything that goes beyond this minimum standard has to be individually achieved, mostly by utilizing market mechanisms (for example, when paying for better education or health care). Those who are unable to employ market mechanisms in order to obtain, for example, adequate education or health care are thrown back to personal resources such as family or church, charity movements, or just suffer bad luck. High degrees of social inequality are generally accepted and justified by equal opportunity and individual performance. In contrast, coordinated welfare regimes like the Netherlands are characterized by a high degree of decommodification, meaning that individual social welfare results not only from individual efforts and socio-economic status, but also from all kinds of redistribution payments. In short, it can be claimed that coordinated production and welfare regimes are characterized by non-fragmentary solidarity among strangers, whereas solidarity and social cohesion are restricted to family or other acquaintances in the liberal regime type (fragmented solidarity).

The approach of eco-cultural adaptation explains these correspondences by arguing that production and welfare regimes coevolve over the course of time and – for that reason – constitute an inherently consistent overall system. In other words, and due to coevolutionary processes of mutual adaptation, welfare services and the prerequisites of the production process are functionally matched and tuned to one another within the liberal and coordinated regime type. For illustration, the nexus between social security, high or low individual investments in firm-specific knowledge, long- or short-term contracts, and the occurrence of incremental or radical innovations is well suited to show how production and welfare arrangements are intermeshed (Estevez-Abe et al. 2001, Iversen/Soskice 2001).

Simplifying this somewhat, a firm's ability to innovate depends on the specific skills of its workforce (or the skill profile of the economy in which the firm is embedded, respectively). The workers' skill profile, in turn, is crucially affected by predominant welfare arrangements. But why should this be so? The central line of argument goes like this: for individual workers, investments in firm- or industry-specific skills are risky endeavors. Specific skills are characterized by low portability (i.e. high asset specificity), meaning that workers commit themselves to specific firms or industries and that their skills (and related investments) are worthless in other contexts. In contrast, general skills "recognized by all employers, carry a value that is independent of the type of firm or industry" (Estevez-Abe et al. 2001: 148). Therefore, and "in the absence of institutional interventions into workers' payoff structure, general rather than asset-specific skill acquisition represents the utility-maximizing strategy" (ibid.: 150). Thus, in coordinated welfare and production regimes, governmental welfare arrangements such as strong dismissal protection or generous unemployment benefits contribute to overcoming employees' inhibitions to invest in specific skills. According to Hall and Soskice (2001), this explains why "*incremental* innovation, marked by continuous but small-scale improvements to existing product lines and production processes" (ibid.: 39) are more likely to occur in coordinated than in liberal regime types.

In contrast, and due to the absence of strong institutional structures, the workforce of liberal economies is predominantly characterized by general skills, meaning that workers can be replaced and substituted much more easily.

> However, the institutional framework of liberal market economies is highly supportive of radical innovations. *Labor markets* with few restrictions on layoffs and high rates of labor mobility mean that companies interested in developing an entirely new product line can hire in personnel with the requisite expertise, knowing they can release them if the project proves unprofitable.
>
> Ibid.: 40

As opposed to incremental innovations, radical innovations are characterized by "substantial shifts in product lines, the development of entirely new goods, or major changes to the production process" (Hall/Soskice 2001: 38 ff.).

This example illustrates not only the extent to which welfare services and the prerequisites of the production process are functionally matched and tuned to one another within liberal and coordinated regime types, but it also suggests that labor and production processes as well as social welfare are subject to similar coordinative principles: centralized and hierarchic coordination combined with long-term planning and strong social cohesion in coordinated paths of adaptation and – on the contrary – market competition, the maximization of individual short-term utility combined with fragmented solidarity in liberal paths of adaptation.

This form of path-internal institutional homogeneity can be explained by the fact that – for reasons of overall societal efficiency as well as institutional and cultural persistency – all realms of social life become subject to one and the same set of coordinative principles in the course of time (cf. subchapter 3.2). Concentrating on path-internal forms of institutional specialization, Schröder (2009) corroborates this central theoretical assumption: "The most important issue is whether there are 'hegemonic belief systems' that not only influence production, but also welfare arrangements and thereby align these to similar principles, causing them to covary" (ibid.: 32). This assumption is based on the idea that normative cultural conceptions such as social justice or economic self-determination "do not single-handedly influence welfare/production constellations, but do shape how social policy-makers think about challenge, crisis and change, meaning certain options are automatically off the table and others are seen to be more legitimate" (ibid.: 32). Related to social learning and cultural evolution as described in subchapter 3.1, the

> mechanism here might be that when faced with challenges, actors look back in their history to what has worked and apply those principles to a new domain. It might therefore be more than a coincidence that, in the US, the welfare system is largely built on the same principles as the production system: a market on which people can assure their well-being by contracting.
>
> Ibid.: 31

But which socio-ecological constellations gave rise to liberal or coordinated forms of institutional specialization? To give the short answer here, the approach of eco-cultural adaptation would argue that cultural preferences for cooperative behavior emerge when the benefits of cooperation outweigh its costs. Otherwise, strong preferences for solitary action are more likely to emerge (subchapter 2.1). Using paradigmatic examples from the Netherlands and the Mississippi delta, this subject will be dealt with in greater detail in subchapter 6.2.

In the meantime, the question of whether the valorization of physical nature is subject to similar path-internal regulatory peculiarities occupies center stage. If this is the case, the valorization of physical nature should follow the same "rules of the game" (i.e. institutional structures and related coordinative principles, North 1990), as can be observed for production processes and the utilization of manpower: centralized and hierarchic coordination combined with long-term planning and the goal of ecological sustainability (analogous to strong social cohesion) in the case of coordinated paths of adaptation as opposed to competitive behavior (for example, market competition) combined with the maximization of individual short-term utility in liberal paths of adaptation.

6.1.2 Corresponding eco-regimes and the valorization of physical nature

Literature problematizing institutional parallels between social and environmental policy is quite scarce (Gough/Meadowcroft 2011). Nevertheless, available studies unanimously agree that social and environmental policies are positively correlated – the stronger social policy efforts, the higher the environmental policy performance, for example, in terms of rigorous or lax environmental legislation (and vice versa). To give an example, and based on quantitative analysis by the Organisation for Economic Co-operation and Development data, Scruggs (1999) finds a robust, positive relationship between corporatist institutions and national environmental performance. Similar to the processes by which they promote economic public goods [...], corporatist institutions also lead to the effective provision of non-economic public goods through their ability to overcome collective action problems that characterize environmental sustainability.

Ibid.: 2

Searching for possible parallels between welfare states and eco-states and showing that both deal with structurally equivalent challenges, Meadowcroft (2005) as well as Gough and Meadowcroft (2011) arrive at similar conclusions, showing that, for example, normative concepts such as justice or equality "can be related to future generations, the distribution of environment-related burdens among rich and poor, and the treatment of nonhuman nature" (Meadowcroft 2005: 9). Finally, and based on such observations John Dryzek (in Gough et al. 2008) tentatively makes the point that "coordinated market economies [...] are better placed to handle the intersection of social policy and CC [climate change, author's note] than the more liberal market economies with more rudimentary welfare states" (ibid.: 336). But why should this be so?

The strong parallels between social and environmental policy are explained by the fact that both realms address similar structural problems, namely the mitigation of negative social and ecological externalities caused by social metabolism in the broadest sense.

Much like the welfare state made externalization of human costs more difficult through legislation and taxation, as well as mitigated the impact of market economy by producing public goods such as health care systems, unemployment support, childcare, and so on, the green state can make exploitation of natural resources more difficult through laws, taxes, and fees, and mitigate environmental impacts through the production of public goods such as public transport systems, green energy systems, habitat preservation and restoration schemes, and sustainable resource utilization.

Duit 2008: 6

And according to Gough and Meadowcroft (2011), there "is growing evidence within the developed world that welfare regimes map on to environmental regimes" (ibid.: 498).[4]

To be more precise, it can be argued that the authors mentioned above conceptualize social welfare (and related goods and services such as education or health care) and environmental goods such as clean air or biodiversity as public goods. In other words, both spheres are shaped by similar structural characteristics and related individual incentive structures regarding their provision and maintenance. Thus, a regulatory inventory designed to cope with social issues should also be applicable to environmental issues. As discussed in Chapter 2, public goods can be described as being non-excludable and non-rival – at least as long as there are no capacity constraints (Olson 1965, McKean 1996, Stavin 2011). Given these crucial characteristics, individual incentives to contribute to the public good are very low, whereas the costs associated with contributing to the public good constitute individual costs, anybody could benefit from these individual investments for free. Thus, there are strong individual incentives for free-riding and shirking, meaning that each individual would be better off if he or she could use the public good without contributing to it which, in turn, results in a situation where each individual is waiting for the others to contribute.

This is why public goods are usually provided by the state, especially – as elaborated above – if characterized by high decommodification and a coordinated production regime. Here, governmental institutions hold strong preferences for the provision of public goods and have enough power to provide and maintain them, for example, by levying taxes (which can be seen as an external force to contribute to the public good from the perspective of individual actors) combined with monitoring and enforcement in case of violation. In liberal welfare and production regimes, in contrast, it can be observed that public goods are turned into marketable goods (i.e. rival and excludable goods) such as gated communities (rather than public security), private schools (rather than public education), or private hospitals (rather than public health care).

These observations can be read as another hint supporting the thesis of path-specific institutional and cultural specialization. It seems that liberal and coordinated welfare and production regimes feature distinct forms of institutional specialization as well as distinct related regulatory approaches and cultural preferences. True to the motto "If you only have a hammer, every problem looks like a nail" (Maslow 1966), the relevant literature obviously agrees that one and the same set of regulatory tools (and related cultural preferences) are applied – regardless of whether social or ecological issues occupy center stage. Accordingly, the coordinated regime type takes recourse to centralized and long-term planning (whereby the general interest outweighs individual short-term interests) in order to do justice to all kinds of social and ecological issues, whereas the liberal regime type fully relies on the allocative efficiency of the market when dealing with social or ecological issues.

This also means that these different regulatory regimes are sensitized to different tasks and challenges, a fact which also goes hand in hand with path-specific blind spots, favorite risks, and related mismatches (cf. subchapter 3.2): whereas the coordinated regime type primarily reacts to issues jeopardizing social and ecological sustainability in the broadest sense, the liberal regime type concentrates on leveled the playing field for competitive behavior and market exchange, especially by guaranteeing property rights, free trade, and market access. Both physical nature and social relationships are subject to short-term individual utility. Governmental regulation or intervention only occurs when market mechanisms are endangered, which may also also include military intervention, as, for example, in the Hindu Kush after the 9/11 attacks or when looting occurred in New Orleans after Hurricane Katrina. Analogously, system-relevant interventions in physical nature rest on the Army Corps of Engineers whose mission it is to deliver "vital public and military engineering services; partnering in peace and war to strengthen our Nation's security, energize the economy and reduce risks from disasters" (US Army Corps of Engineers 2014).

6.1.3 Summary

Recalling the initial aspiration of the present chapter – namely to describe and explain how and why ecological and social challenges such as flooding or social inequality are dealt with differently along the Rhine–Meuse–Scheldt and Mississippi deltas – it was assumed that such differences could be ascribed to distinct forms of path-specific institutional specialization and related cultural preferences. According to the views expressed in the previous section, this assumption appears to hold true. Bringing together the approach of eco-cultural adaptation with the relevant literature discussing path-internal elective affinities between regulatory approaches coordinating the utilization of manpower and physical nature, it was possible to identify path-specific forms of institutional specialization. Coordinated regime types are specialized in the provision of public goods and related regulatory principles (especially long-term planning and centralized coordination combined with strong social cohesion), whereas liberally biased regimes primarily trust in the coordinative abilities of the market as spelled out in neoclassic textbooks. In simple terms, and in diametric contrast to the coordinated regime type, this means that the liberal regime type is marked by institutional environments fostering competitive behavior and the provision and distribution of excludable and rival goods and services. Apart from that, non-regulation or laissez faire-like conditions are preferred. Consequently, and because flood protection and social welfare fall into the category of public goods rather than private goods, liberally biased regimes arguably suffer from structural disadvantages. However, these structural prerequisites are only disadvantageous from the perspective of an external observer interested in the gearing of physical nature, technologies, institutions, and cultural preferences. This does not mean that local actors necessarily have to share this view. On the contrary, it could be argued that the liberal bias is accompanied by ideational convictions (as, for example, embodied in the "rags to rich tale" or social-darwinistic "bootstrap individualism") legitimizing or even idealizing occurring mismatches. From an external perspective, trying to provide public goods such as flood protection in liberally biased environments is arguably doomed to failure. Or to put it in other words: the sheer usage of similar techniques such as levees does not necessarily result in comparable degrees of flood protection, both in quantitative and qualitative terms. As will be shown below, it does make a considerable difference whether levees are provided under the condition of centralized coordination, strong social cohesion, and third-party enforcement or whether their implementation is attempted in an institutional environment specialized in fostering competitive behavior and the maximization of individual short-term utility.

6.2 Case studies

As shown in the previous chapter, the Rhine–Meuse–Scheldt and Mississippi deltas are subject to different regulatory environments organizing the utilization of physical nature and society in distinct ways. On the basis of these considerations, this section is devoted to the following questions: what can be said about the emergence of these path-specific institutional peculiarities? How did they mold past and still mold present approaches to ecological and social challenges? And – last but not least – how do they affect individual resources to cope with extreme weather events such as Hurricane Katrina?

Trying to answer these questions, the subsequent arguments are based on the following assumptions: in general, early settlers in the Netherlands and in North America faced one and the same challenge – namely the development of habitable space. However, they were confronted with very different situations. While the Dutch had to make themselves at home in a rather homogeneous physical environment primarily characterized by marshy landscapes, spatial scarcity, land subsidence, and the permanent threat of being flooded, the first settlers in North America faced the challenge of colonizing a vast landmass characterized by a multitude of different types of physical nature (for example, with regard to climate zones). While the marshy landscapes along the Rhine–Meuse–Scheldt delta selected for cooperative behavior and related technologies, the North American frontier primarily favored technologies (and respective behavioral strategies) that could be individually and spontaneously applied. Accordingly, and in the course of time, appropriate institutional structures and cultural biases emerged, either emphasizing hierarchic coordination as well as social and ecological sustainability in the Netherlands or solitary action, strong individualism, and rather laissez faire-like modes of molding and exploiting physical nature and society (and social relationships, respectively) in the United States. Up to the present time, the frontier ideology finds expression in strong individualism, (market) competition, a low level of decommodification, and related legitimizing social myths (for example, "rugged individualism", the "up-by-the-bootstraps philosophy", and the "rags to riches" tale).

Generally speaking, it can be said that appropriate eco-cultural fabrics emerged both in the Netherlands and in North America over the course of time. However, and as briefly illustrated by the example of the Dust Bowl in subchapter 3.2, a liberally biased mode of life tends to produce severe mismatches and ecological crises when faced with sensitive ecosystems. In other words, it is claimed that a liberally biased mode of life and related eco-cultural matches perfectly fit societal and ecological constellations with low payoffs to (large-scale) cooperation, but result in severe mismatches when superimposed on sensitive ecosystems – but that is exactly what happened along the Mississippi delta and in New Orleans. To put it even more pointedly: while appropriate matches between physical nature, technologies, institutions, and cultural preferences endogenously coevolved along the Rhine–Meuse–Scheldt delta over the course of time, the liberally biased mode of life – perfectly suited to individually taming the frontier – was exogenously superimposed onto the marshy wetlands along the Mississippi delta and – since then – has resulted in severe tensions between liberally biased institutions and cultural preferences on the one side and the functional necessities associated with the provision of public goods such as flood protection and social security on the other.

The discussion evolves as follows: in a first step, and in order to provide a foil for comparison with flood protection and social welfare along the Mississippi delta, the formation of the Dutch adaptational path is described from a rather eco-deterministic perspective. Second, and from a rather cultural-deterministic perspective, it is shown what happens when a liberally biased mode of life coincides with a physical environment which – as is the case in the Mississippi delta – demands centralized coordination and cooperation rather than solitary action.

Before going into detail, a short clarification might be appropriate here: the author does not believe that specialized and capital-intensive hydraulic engineering is the only possible way to ensure the habitation or utilization of deltaic wetlands. This might only hold true when permanent settlement, intensive land use, and industrialization are the declared goals. Otherwise, and as the example of floating gardens in Bangladesh impressively shows, living in deltaic wetlands can also be sustained by nomadic lifeforms combined with low-tech adaptational measures (spontaneously and individually applicable) and limited solidarity based on kinship or clan.[5] In short, one should not get the impression that there is an automatism directly linking deltaic wetlands to large-scale hydraulic works as observed along the Rhine–Meuse–Scheldt and Mississippi deltas.

6.2.1 The Dutch path of cooperative dike-building

The following analysis basically rests on the extensive work of Yda Schreuder (2001), Martin Reuss (2002), William TeBrake (1995, 2002), Petra van Dam (2002), Arne Kaijser (2002), Harry Lintsen (2002), Wiebe Bijker (2002), and Gerard van de Ven (2004). It is used as empirical material and – for that reason – quoted at length in some places. Trying to summarize the general thrust of their work, it can be said that all of them – from different perspectives – shed some light on the dynamic relationship between dynamic physical nature, changing technologies, and the emergence of the respective institutional structures and cultural preferences. Thereby, the authors listed above agree that fundamental cornerstones of the current Dutch political and economic system have their origin in hydraulic engineering and related organizational and institutional necessities.

Starting their analysis in the thirteenth century, the authors are able to show how the Dutch path of eco-cultural adaptation became more and more solidified over the course of time. This process could be described as either a self-stabilizing circle or a technological and cultural lock-in: implementing new technologies such as simple drainage canals resulted in land subsidence (changes in physical nature) which, in turn, demanded new technologies such as levees, sluices, or wind-driven water pumps. Implementing these new technologies again accelerated the process of land subsidence (Kaijser 2002, van Dam 2002). Thereby, the authors cited above show how the dynamic interplay of (technologically induced) environmental changes and the implementation of related technological coping strategies regularly entailed the emergence of appropriate institutional structures (and cultural preferences) coordinating the task of cooperating in hydraulic engineering on different spatial scales (van Dam 2002).

The beginnings of hydraulic engineering along the Rhine–Meuse–Scheldt delta go back to individual and uncoordinated attempts to gain farmland by draining peat bogs.

> While dikes, dams, sluices, and drainage canals had become necessary for human occupation by 1300, this had not always been the case. Until the middle of the twelfth century, settlement and agricultural use of the lowland zone were made possible by straightforward drainage, using simple techniques – digging small, shallow ditches to enhance the normal flow of water. Unfortunately, this initial round of drainage had a number of unintended consequences, the most important of which was the subsidence of the drained lands. These became so susceptible to flooding as to cause what can only be described as an environmental crisis. A *waterwolf* stalked the land.
>
> TeBrake 2002: 476[6]

In response to an increasing number of flood events, more complex and sophisticated drainage systems, covering more and more land and involving more and more actors, were established.

> In short, residents of the lowland zone now had to measure, plan, and execute with meticulous attention to detail. The simple digging of ditches designed to enhance the normal flow of water from higher to lower elevations had been replaced by real hydraulic engineering.
>
> TeBrake 2002: 486

From the perspective of eco-cultural coevolution, it can be argued that this historic course of events crucially influenced the emergence of an adaptational habitat primarily marked by hierarchic coordination and long-term cooperation. As argued in Chapter 2, the relationship between costs and benefits associated with cooperation determines whether cooperative behavior or solitary action will prevail. Cooperative action will emerge whenever the benefits of cooperation outweigh its costs, whereas solitary action will prevail whenever the costs of cooperation outweigh its advantages. Given the high cost of coordinating cooperation, for example, in terms of negotiation and monitoring costs in the case of simple farming and drainage endeavors, it made absolute sense for early settlers not to cooperate (Ellickson 1991, 1993). In retrospect, however, and as described by TeBrake (2002) or van de Ven (2004), it can be argued that the (long-term) costs associated with uncoordinated drainage and farming endeavors – particularly in terms of land subsidence, land loss, salinization of groundwater and soil, and casualties due to flooding[7] – clearly exceeded those of coordinating cooperative action. In response to these man-made changes in physical nature, those affected by land subsidence and flooding began to experiment with new technologies, especially with levees and sluices.

As a result of the dynamic and intricate characters of coastal marshlands and fluvial ecosystems (TeBrake 2002, van de Ven 2004, Hudson et al. 2008, Syvitski et al. 2009), the introduction of technologies such as levees and sluices presents a necessary yet insufficient solution to sustainable water management. Implementing such technologies in an uncoordinated way at the local level, neighboring communities – as an unintended side effect – easily harmed each other, for example, by redirecting floods to each other's territories or altering groundwater levels. As a result of restricted local resources – especially with regard to labor, capital, and knowledge – the first attempts at hydraulic engineering often did not yield the anticipated results, meaning that flooding nevertheless occurred. The situation gradually improved with the successive establishment of local water authorities, bundling the relevant know-how and orchestrating cooperation.

> In the thirteenth and fourteenth centuries the establishment of regional water authorities greatly increased the ability of the Dutch to cope with flooding and improve drainage. But a paradox developed: the more successfully they drained the land and created favorable conditions for agriculture in the short term, the faster the former peat bogs sank, threatening the conditions for agriculture in the long term. Short-term success thus led to long-term crisis.
>
> Kaijser 2002: 529

Thus, inhabitants of the lowland zone had to understand

> that land drainage was not a one-time achievement but something that they had to continue doing, or the original, waterlogged conditions inevitably would return. [...] Finally, they soon learned to cooperate with each other, for one person's drainage all too easily became the next person's floodwater.
>
> TeBrake 2002: 490

In other words, inhabitants of the lowland zones not only had to engineer proper levees and sluices, but also had to develop appropriate institutional structures lubricating these technologies (i.e. institutional structures able to cope with problems of collective action), especially by defining, controlling, and enforcing crucial rights and obligations of all parties involved (North 1990, Kaijser 2002).

The most prominent features of hydraulic works and flood protection are non-excludability and high payoffs to cooperation with regard to their provision and maintenance (McKean 1996, Stavin 2011). In principle, and due to problems of collective action such as free-riding and shirking, this constellation makes cooperation a rather unlikely scenario (Olson 1965, Poteete et al. 2010). At the same time, however, cooperation is not only required by the mutual dependencies of neighboring communities, but presents the only way to exploit economies of scale resulting from the use of common infrastructure. Despite looming problems of collective action, the Dutch apparently worked out how to provide stable and durable living conditions in deltaic wetlands. But how can their success be explained?

Their key to success lies in particular organizational and institutional structures which coevolved with hydraulic engineering techniques. "Many of these devices were large, and required the cooperation of many people to build. Such collective action in turn depended on institutions that could define a division of tasks and responsibilities and set sanctions for misconduct" (Kaijser 2002: 522). In other words, and to avoid dilemmas of collective action and to allow for cooperation across time and space (for example, between different communities and intergenerationally), such organizational and institutional structures not only had to define task-specific rights and obligations for all the actors involved, but also to allow for control and enforcement (North 1990, Ellickson 1993).

> The earliest of these, with primarily local responsibilities, predated the emergence of anything resembling a state. From the twelfth century onward, many such local, collective units joined together, as cells do, to form higher-order units for common purpose [...]. By 1350, a series of autonomous regional water boards had developed that managed drainage and flood control in most of the lowland zone [...]
>
> TeBrake 2002: 490

The specific functioning of these early regional organizational structures is summarized by Kaijser (2002) in the following way:

> The early dikes and drainage canals were constructed within the existing political framework, with the village council as a key organization. The village council made agreements with neighboring villages regarding the division of responsibilities for maintaining the larger hydraulic structures. It also formulated operational rules, called *keuren*, concerning the maintenance of the hydraulic structures within its boundaries, specifying the requirements that had to be met and the sanctions in case of violation. Every farmer was given responsibility for maintaining a part of a dike or a canal in proportion to the size of his farm. A special inspection committee was chosen from among the farmers each year to monitor the maintenance work, and this monitoring was done very carefully, particularly for the dikes, as it was well understood that a dike is like a chain – it is only as strong as its weakest link. The committee conducted an inspection (*schouw*) three times a year, and the sanctions for deficiencies increased at the second and third inspection. This careful monitoring gave all farmers two kinds of incentives for fulfilling their maintenance work: they wanted to avoid criticism and sanctions for shortcomings in their own efforts, and they were convinced that all other farmers were forced to fulfill their tasks as well, which made the whole enterprise worthwhile. Thus these rules effectively precluded free-rider behaviour.
>
> Kaijser 2002: 526

In this context, the elaborations of van Dam (2002) deliver further valuable information about the organizational properties of local water authorities, for example, regarding their far-reaching powers such as jurisdiction and levying taxes, and provide deeper insights regarding the question of how the coordination of hydraulic infrastructure – that is, cooperation, mutual control, and sanctioning – were embedded into daily routines and practices.

> Beginning in the tenth and eleventh centuries the regional water authorities came to serve as courts of justice, with jurisdiction only over regional hydraulic works. On designated days, local water authorities sent representatives to perform maintenance at regional structures or to supervise construction by entrepreneurs, who might not be resident within the boundaries. In order to pay their representatives and these entrepreneurs, local communities levied taxes on landed property. The main task of the regional water authorities was to inspect the works, a task performed during ritualized meetings at fixed dates associated with traditional seasons of the agricultural calendar. The aim was to ensure that maintenance was completed before the autumn storms brought high floods, and social and institutional pressure usually sufficed to ensure a satisfactory level of performance. If problems developed, the officer of the regional water authorities would simply stay in the local inn and hold court until the maintenance work was finished. The penalties imposed on local farmers who failed to pass these inspections included fines, which were not only an economic burden but also a source of shame, for days there would be jokes about it, and people would remember very well who did not repair his dike in time.
>
> van Dam 2002: 502 ff.

From a theoretical perspective, and according to North (1990), problems of collective action and related agency costs are mainly rooted in information asymmetries among the stakeholders concerned. By implication, this means that cooperation can be facilitated by institutional structures allowing for certain degrees of transparency as well as mutual controllability and sanctioning. Accordingly, it does not come as a surprise that cooperation was based on highly personalized and frequent interactions at the local level in combination with third-party enforcement embodied by (transregional) water authorities.[8] Because maintenance work was mostly carried out by local (and neighboring) farmers bound to the soil – meaning that they had to get along well with each other over generations – each of them had strong incentives to invest in their good reputation and to fulfill their duties (North 1990, Henrich et al. 2004). Based on both spatial proximity and frequent interaction – which were typical attributes of rural life in the Netherlands in those days – mutual control and sanctioning was more or less embedded in daily routines and practices. According to North (1990: 34 ff.), these conditions should suffice to enable cooperation on the local level. However, and because riverbanks, shorelines, or penetrating water do not stop at municipal borders, some sort of trustworthy third-party enforcement had to be established to enable intercommunal cooperation – that is, cooperation amongst more or less foreign actors who, in contrast to neighbors, defy mutual reputation building as well as direct control and sanctioning on a daily basis. In addition, different local perspectives, rules, and technological traditions had to be coordinated on a transregional level. Thus, water authorities with far-reaching powers (especially regarding jurisdiction and levying taxes) and inspection committees with elected and rotating staff (guaranteeing that everybody involved would not only have the role of an appraisee, but eventually would also officiate as an inspector) emerged to enable intercommunal cooperation – that is, cooperation transcending time and space – via hierarchic coordination, monitoring, and enforcement.

Basically, and from a structural perspective, this form of hierarchic coordination combined with local responsibilities has not changed up to the present time, even though a trend toward more centralized coordination can be observed (van Dam 2002). Industrialization and rising degrees of labor division, urbanization, and related phenomena – such as rural flight and the proletarianization of former peasants – put the Dutch path of eco-cultural adaptation under severe pressure. The rising energy demands of the cities and the emerging industrial production regime were met by peat mining at the periphery and resulted in accelerated land subsidence and the unintended emergence of lakes. In addition, intensification and commercialization of agriculture further contributed to soil compaction. As a result of rural flight, the

> traditional method of maintaining dikes – using the labor of farmers – gradually gave way to a system of professional maintenance financed by taxes [...]. An army of dike workers recruited from the rural proletariat could now be mobilized to improve and maintain the dikes, and highly skilled hydraulic engineers supervised their work.
>
> Kaijser 2002: 541 ff.

Thus, the accomplishment of hydraulic engineering not only had to be revised with regard to the labor force, but also had to allow for the newly emerging demands associated with industrial production regimes, for example, rising energy demands and the transportation of raw materials or commodities. Canals no longer exclusively served the purpose of drainage but increasingly gained importance for transportation and shipping. In turn, this presupposed the definition and implementation of nationwide standards regarding the breadth and depth of canals and sluices as well as navigable water levels. In short, hydraulic engineering had to account for an ever-increasing amount of varying demands which, in turn, had to be synchronized and coordinated on a nationwide level. Therefore, local and regional water authorities were supplemented by another more centralized organizational structure: since 1798, the governmental Rijkswaterstaat has coordinated hydraulic engineering. Nevertheless, local water authorities still play a major role even though in reduced absolute numbers (Bijker 2002, Kaijser 2002, Lintsen 2002).

It should have become obvious to the reader by now that hierarchic coordination, cooperation between strangers, a fair distribution of (financial) burdens, strong social cohesion, as well as balancing the adverse effects between short-term interests (for example, individual profits gained from peat digging) and the general public good (especially long-term flood protection) are fundamental constituents of the Dutch eco-cultural path of adaptation.[9] Moreover, it can be claimed that these constituents are deeply rooted in the historical experience that the Dutch "man-made lowlands" (van de Ven 2004) are very sensitive to changes in physical nature and society, resulting in the crucial insight that ecological planning and sustainability are a means to the end of social security – and not two opposing goals. Thus, durable stabilization of the fragile balance between physical nature and society, as well as managing the unintended effects of their dynamic interplay, has always been a central challenge the Dutch path of eco-cultural adaptation had to face and which has significantly shaped its current form. Thereby, hierarchic coordination and regulation, long-term cooperation, far-sighted planning, as well as precaution (rather than aftercare) significantly contribute to this balancing act.

In the course of time, these institutional structures and cultural preferences have become more and more detached from their functional origins and permeated ever more realms of society. In other words, strong cultural preferences for cooperative and collective action turned into hegemonial guiding principles of social action. In this context, Schreuder (2001) points out that based "on a tradition of cooperation, census building, and democratic self-rule, the Dutch have revitalized a corporatist approach to economic and environmental planning" (ibid.: 237). In the same vein, Kaijser (2002) states that the "strong tradition of trying to find solutions that are acceptable to all parties involved [...] has been transferred from water institutions to the general political culture of the country" (ibid.: 547). These observations can be seen as another hint corroborating the findings of subchapter 6.1, according to which eco-cultural habitats of adaptation are markedly affected by path-internal institutional specialization.

Summing up, and in order to provide a foil for comparison with flood protection and social welfare along the Mississippi delta, an attempt has been made to understand how the Dutch habitat of eco-cultural adaptation evolved in the course of time from an eco-deterministic perspective. Elaborating how public goods can be provided and maintained despite problems of collective action, the extent to which the provision of public goods (such as flood protection and related technological devices) depends on an appropriate institutional environment, as well as corresponding cultural preferences, has been illustrated. In short: without appropriate institutional embeddedness, technological devices such as levees or sluices probably yield only suboptimal performance or may even be worthless.

With regard to the central question of this chapter, the empirical evidence presented and discussed above suggests that path-specific institutional specialization and corresponding cultural preferences are the reasons why social and ecological tasks and challenges are similarly approached within one and the same habitat of eco-cultural adaptation. In other words, each type of path-specific institutional specialization is accompanied by particular regulatory approaches organizing the utilization of physical nature and society. Because the institutional analysis of subchapter 6.1 and the findings of the present coevolutionary-historic reconstruction of the Dutch habitat of adaptation come to similar conclusions (despite different analytical perspectives and empirical material), it can be argued that a robust picture of the Dutch eco-cultural path of adaptation can now be drawn. Armed with these findings, the Mississippi delta and New Orleans will be addressed next.

6.2.2 Hurricane Katrina and the liberal approach to flood protection in the United States

It has been argued that appropriate matches between physical nature, technologies, institutions, and cultural preferences endogenously coevolved along the Rhine–Meuse–Scheldt delta over the course of time. Taking these considerations as a foil for comparison, the present section deals with the contrary constellation: in the case of the Mississippi delta, a preexisting liberally biased mode of life – perfectly suited for individually taming the frontier and colonizing the West – was exogenously applied to the deltaic wetlands. Ever since, this constellation has resulted in more or less severe tensions between the functional prerequisites of public goods on the one hand and cultural preferences emphasizing solitary action, the short-term utilization of individual utility, and the application of unconnected technologies in a rather laissez faire-like institutional environment on the other.

This perspective is similar to Worster's (1979) explanation of the Dust Bowl in the Great Plains (cf. subchapter 3.2). Basically, he ascribes the Dust Bowl to a crucial maladjustment between a sensitive ecosystem (the semiarid Great Plains) on the one side and cultural values "taught by capitalism" on the other. The Dust Bowl, Worster argues,

> cannot be blamed on illiteracy or overpopulation or social disorder. It came about because the culture was operating in precisely the way it was supposed to. Americans blazed their way across a richly endowed continent with a ruthless, devastating efficiency unmatched by any people anywhere. When the white men came to the plains, they talked expansively of "bursting" and "breaking" the land. And that is exactly what they did. Some environmental catastrophes are nature's work, others are the slowly accumulating effects of ignorance or poverty. The Dust Bowl, in contrast, was the inevitable outcome of a culture that deliberately, self-consciously, set itself that task of dominating and exploiting the land for all it was worth.
>
> Ibid.: 4

Transferring this perspective to New Orleans, attempts have been made to explain past and present struggles to provide and maintain public infrastructure such as sewage canals or levees (and related goods such as public health and flood protection) by suboptimal matches between deltaic wetlands, hydraulic works, and a liberally biased "toolbox" containing suboptimal coordinative instruments and cultural preferences to provide these goods. With this, the present section revolves around the crucial question of how a liberally biased mode of life guides the course of social action and – in doing so – crucially affects humans' interferences with physical nature and each other.

But where did the North American liberally biased cultural preferences initially come from? As indicated by Frederick Turner's (1894) Frontier Thesis, it can be argued that European settlers – leaving their religiously oppressive, highly regulated, and often (quasi) feudalistic home countries behind – especially emphasized strong property rights as well as individual freedom (including political autonomy and religious freedom) when colonizing their new homeland. Under the impression of feudalistic European serfdom, individual freedom and political autonomy were inextricably linked with property in land. As mentioned earlier, this fundamental attitude is also reflected in the Homestead Acts, which granted parcels of land (usually 160 acres) to individual homesteaders at little or even no cost (Elkins/MacKitrick 1993, Foner 1995). In a happy coincidence, the combination of individual freedom, strong property rights, and comparably weak coordinative frameworks was generally well suited to conquering the frontier, particularly to breaking the uncultivated but arable ground (Turner 1894, Ellickson 1993).

As summarized by Worster (1979), these cultural preferences are accompanied by the following ecological values regulating the valorization of physical nature:

1. *Nature must be seen as capital.* It is a set of economic assets that can become a source of profit or advantage, a means to take more wealth. Trees, wildlife, minerals, water and the soil are all commodities than [sic] can either be developed or carried as they are to the marketplace. A business culture attaches no other values to nature than this; the nonhuman world is desanctified and demystified as a consequence. Its functional interdependencies are also discounted in the economic calculus.
2. Man has a right, even an obligation, to use his capital for constant advancement. [...]
3. The social order should permit and encourage this continual increase of personal wealth. [...]

   The notion that nature puts restraints on what man can do in those businesses was as abhorrent to him as were social control.

Ibid.: 6

Without going into further detail here, and trying to intertwine these considerations with the findings of Chapter 3 (cf. Table 3.1) and subchapter 6.1, the most central characteristics of the liberal bias can be summarized in the following way: liberal cultural paths initially originate from cooperative disadvantages which foster solitary action and entail the emergence of cultural preferences emphasizing strong individualism, individual autonomy, and freedom. In the same way, socially unconnected technologies which can be applied individually and spontaneously are preferred. Solidarity is restricted to kin, clan, or ethnicity. As a result of low cooperative advantages and low payoffs to reputation building, time preferences range at a comparably low level. Thus, humans' interference with physical nature and each other are utilized according to individual short-term interests. Apart from securing property rights and some other crucial prerequisites for solitary action, competitive behavior, and market exchange (for example, standardization), the regulatory and institutional environment of liberal habitats is rather weak.

Taken together, liberal eco-cultural habitats are especially predestined to deal with socio-ecological challenges which concern individual actors (or small collectives such as local populations or single firms) and demand for decentralized, spontaneous, and unbureaucratic actions, decisions, and related technologies within short- or medium-term planning horizons. Or to put it the other way round: liberal habitats are particularly vulnerable to socio-ecological challenges which go beyond individual capabilities and demand for long-term planning and hierarchic coordination. Consequently, and in sharp contrast to the overall situation in the Netherlands, public goods such as flood protection, public security, or social welfare are only sparsely provided (if at all), which, in turn, carries new vulnerabilities in its own right, especially with regard to preparedness for flooding and – due to high levels of social inequality – negatively affected individual coping capacities in case of calamity.

In other words, and in contrast to the Dutch, it could be concluded that the inhabitants of New Orleans are vulnerable to flooding in a twofold way: first, as a result of the inadequate institutional embeddedness of hydraulic works, it can be assumed that the general level of preparedness is suboptimal, meaning that flooding becomes more likely to happen. Second, as a result of low degrees of decommodification and high degrees of social inequality, individual capacities to cope with and recover from flooding are unequally distributed. As will become obvious below, these assumptions are not only of a theoretical nature, but can be substantiated by empirical evidence illustrating the tension between a liberally biased mode of life and the functional prerequisites of public goods. Thereby, the intricate interplay between socio-economic status, elevation profile, and the spatially unequal availability of public goods in past and present New Orleans is of special interest.

New Orleans can be described as a bowl-like city located (mainly) between Lake Pontchartrain to the north and the Mississippi river to the south. Apart from natural levees along the riverbanks, major parts of the city lie below sea level (Hiles 2007, Hudson et al. 2008). As a result of these geomorphological characteristics, New Orleans is especially prone to flooding caused by high water levels in the Mississippi river or Lake Pontchartrain (due either to strong precipitation in the upper reaches of the river or by flood waves built up by strong winds). Even if the levees resist high water levels, flooding caused by strong (subtropical) precipitation and exceeding the short-term capacity of water pumps can nevertheless occur. Similar to the situation in the Netherlands, New Orleans is caught in a vicious circle: over the course of time, drainage endeavors resulted in higher population densities which, in turn, went hand in hand with land subsidence and surface sealing which – again – demanded better adapted drainage systems, and so on (Colten

2000, 2005). Here again, short-term success resulted in long-term problems: the closer the levees bordering the river on both sides approach (meaning that the riverbed becomes increasingly narrow) and the more land that is drained, compacted, and sealed, the less water can be absorbed by the sponge-like marshy wetlands. In addition, and as a result of human interference with the deltaic wetlands which, in turn, mostly followed economic rather than ecological premises – the formerly marshy wetlands can no longer serve as natural wind- and wave-breakers, meaning that both wind and water hit the city at full force (Hiles 2007, Colten 2009). Consequently, flood waves hit new highs which, in turn, demands reinforced and higher levees and stronger water pumps to keep the bowl dry and safe.

In contrast to the Netherlands, however, these intricate geomorphological and climatic conditions were not dealt with in an egalitarian way (true to the motto "one for all, all for one"), but rather in an individualistic and anti-egalitarian manner combined with half-hearted governmental attempts to provide the most crucial public goods. Craig Colten (2005) puts it like this:

> In the western states, there is a common saying that water flows toward money. In water-scarce areas, those who can afford to secure water rights are able to direct the precious fluid towards their property. In the Crescent City [i.e. New Orleans, author's note] the situation is reversed. Built entirely on an alluvial floodplain, subsequently surrounded by levees to keep out floods from the Mississippi River, and now sinking under its own weight, the city and its suburbs must collect and pump out an average 60 inches of rain that falls annually into what is often described as a giant bowl. All residential or commercial construction within the basin sheds rainfall, adding to surface runoff, and is subject to periodic flooding from storms that exceed the drainage system's short-term capacity. The city and its occupants have situated themselves in harm's way. Within the general pattern of flood risk, water flows away from money – that is: away from the property of those who can afford to live in less flood-prone areas and those with the influence to secure adequate publicly financed water-removal services.
>
> Ibid.: 141

Water flows away from money. This simple sentence impressively encapsulates the historically grown and intricate interplay between socio-economic status, power, residential area within the bowl (and its elevation above sea level), and the spatially unequal availability of public infrastructure such as flood protection devices. In more general terms, and bearing in mind Colten's (2005) elaborations regarding the past and present provision of public goods in New Orleans, the more general statement "harm flows away from money" might be even more appropriate. To understand why, it is important to take a closer look at the spatial correspondences between New Orleans' social geography and topography.[10]

As a rule of thumb, and with regard to individual households, it becomes apparent that socio-economic status, place of residence, and elevation above sea level are positively correlated (apart from some neighborhoods in the northeastern parts of the city, for example, Lakeview). Whereas the white middle class predominantly resides on the natural elevations along the riverbanks, African Americans (as well as Hispanics or Asians) mainly live in low-lying areas (Cole et al. 2007). Such distribution is not new but is a long-standing characteristic of New Orleans. This form of spatial and racial segregation is due not only to different land prices and levels of rent, but can also be explained by everyday racism. But there is more to it than that: as Colten's (2005) historical discussion with regard to the provision of public infrastructure elucidates, public infrastructure was and still is provided in spatially unequal ways, privileging commercial districts and residential areas of the white and influential middle class over low-lying areas predominantly inhabited by socially deprived African Americans (Colten 2005: 77 ff.).[11]

What does this actually mean? Whereas the predominantly white middle-class lives on heightened grounds and – in addition – has the financial and political means to improve their living conditions by either private foresight (for example, by building their homes on stilts or by buying flood insurance) or by lobbying for publicly financed infrastructure, socially deprived households have to live in areas which are more likely to be flooded and – at the same time – are only poorly equipped with public infrastructure. Moreover, these households have fewer or no resources at their disposal to privately adapt to these unfavorable natural and man-made living conditions. Especially during the nineteenth century, and due to the bowl-like character of the city, spatial and racial segregation combined with unequally distributed public infrastructure resulted in a situation in which low-lying areas were not only negatively affected by flooding, but – more generally – by all kinds of fluid negative externalities, such as feces and sewage (from private homes as well as from factory installations such as slaughterhouses). Moreover, and in order to gain habitable space, low-lying areas had often been filled with all kinds of trash and debris contaminating the air, soil, and groundwater.

Consequently, the inhabitants of these low-lying areas were severely affected by harmful airborne effluvia, polluted water, and epidemics such as yellow fever transmitted by mosquitoes which found optimal growing conditions in the shallow wastewater pools and the humid subtropical climate (Colten 2005: 47 ff.). In contrast, "residents with means took flight from the summer pestilence and avoided the most serious epidemics" (ibid.: 43).

> Throughout the early nineteenth century, the city had been unable to pass the costs of drainage on to others [...]. This cost weighed disproportionately on the poor in [the] form of huge death tolls during the frequent yellow fever outbreaks. Wealthy citizens paid for inadequate drainage through the expense of taking refuge in the country.
>
> Ibid.: 44

Although this situation gradually improved throughout the twentieth century, Hurricane Katrina starkly revealed that spatial segregation, social inequality, and the unequal availability of public infrastructure – especially hydraulic works – remain a serious issue. Malicious tongues might even state that New Orleans – throughout its history – has earned a reputation for dumping its trash where the bowl is deepest, whereby no difference was (and partially still is) made between sewage, floodwater, debris, trash, or unwanted fellow men such as socially deprived households.

Given this symptomatic description highlighting some manifest consequences of a liberally biased mode of life in New Orleans, an attempt is made now to develop a deeper understanding of the underlying eco-cultural "mechanics" that could give rise to this socio-ecological mélange. Based on Colten's (2005) historical analysis, the following general pattern seems to apply to the questions of when and under what circumstances public infrastructure and related goods were provided in the past: fairly symptomatic of liberally biased paths of adaptation, public goods were predominantly thematized in terms of nuisance – that is, when negative externalities infringed on the enjoyment of one's property or harmed personal health (Colten 2005: 47) – or when negative externalities endangered commercial activities. However, private lawsuits are reserved for those who can afford them. In response, and due to the fact that nuisance mostly affected all of the densely populated city (and not just single individuals who could bargain individual solutions), governmental bodies usually issued legislative regulations to bring about relief. Their implementation, however, was mostly left to individual landowners or sourced out to private companies.

As a result, and because individual resources and the landowners' willingness to contribute to the public infrastructure (or to pay for its implementation) varied significantly, flood protection or sewage systems were often piecemeal. In New Orleans this resulted not only in a situation where the wealthy were better supplied than the poor, but also in economic inefficiencies: although investments were made, positive effects of scale remained mostly untouched and damage occurred nevertheless. Unlike in the Netherlands, Colten (2005) repeatedly explains the poor implementation of public infrastructure with the city's quasi non-existent enforcement policies. Thus, the regionwide implementation of adequate public infrastructure in particular suffered from insufficient centralization, coordination, and enforcement – that is, insufficient institutional embeddedness of connected technologies – combined with political and individual opportunism as well as administrative and social fragmentation.[12]

In this context, it is also important to bear in mind that the fragmentary distribution of public infrastructure favoring the wealthy was not necessarily perceived as unjust by social actors. Because the liberal bias is characterized by the strong belief in equal opportunities and justice of performance combined with strong individualism, it could even be argued that this situation was just fine: everybody got what he or she deserved. Those who do not want to live in the sewer just have to pull themselves up by their own bootstraps. As will shortly become apparent, businessmen apparently only lobbied for relief for the poor (for example, in terms of sewage systems) once persistent health issues seriously threatened their business activities.

In the following, selected historical examples highlighting crucial aspects of this general pattern describing the provision of public infrastructure and related goods in New Orleans are reported. Thereby, public health and flood protection as well as related (connected) technologies such as sewage systems, fresh water supply, and hydraulic works are of specific interest. The first examples stem from the realms of flood protection and illustrate what is meant by insufficient institutional embeddedness as well as political and social fragmentation.

In the eighteenth century, governmental bodies demanded individual landowners living along the riverbanks in the agricultural hinterlands of New Orleans to build levees and – in doing so – to contribute to New Orleans' protection and economic success.[13]

> Levee construction became a sizeable investment for landowners and was only feasible for wealthy planters using slave labor. Consequently, most small landholders were unable to complete their protective structure. Although even wealthy planters may have been reluctant to make such large investments, the threat of confiscation for failure to comply motivated most to participate – to some extent. [...]. Despite a sound policy, privately built structures were notoriously inconsistent in design and effectiveness, and floods continued to breach these ever-lengthening earthen embankments.
>
> Colten 2005: 20

A similar example illustrating the uncoordinated and inconsistent approaches to flood protection dates back to 1836:

> During a period of relative flood security in New Orleans, political fragmentation within the city prompted the creation of three separate municipalities with one mayor. Beginning in 1836, each municipality had responsibility for maintaining its own levees along the urbanized waterfront. Since the three municipalities differed greatly in terms of their tax base, levee maintenance was uneven.
>
> Ibid.: 25

Given these examples, and remembering the highly coordinated, coherent, and specialized approaches to flood protection in the Netherlands, it comes as no surprise that flood protection along the Mississippi delta only yielded suboptimal results.

Another example illustrating the liberally biased mode of life and its mode of operation can be seen in the interplay of flood protection and social fragmentation. Whereas the overall situation in the Netherlands was characterized by strong social cohesion – true to the motto: one for all, all for one – spatial and racial segregation in New Orleans can be seen as an indicator of fragmented solidarity. This becomes observable on a smaller as well as on a larger spatial scale. In the same way as the influential middle class accepted the repeated flooding of low-income quarters and more or less unintentionally used them as cesspools,

New Orleans' business community – when faced with the Mississippi river flood of 1927 – successfully lobbied to sacrifice the peripheral St. Bernhard Parish (situated in downstream New Orleans) as a retention area to relieve the city's levee system and to save their marketplace (Gomez 2000). When rumor spread that St. Bernhard might be sacrificed to save the affluent center, armed levee guards appeared along St. Bernhard's riverbanks. Whereas the Dutch appointed dike watches around the clock to assess the technical conditions of hydraulic structures in case of emergency, inhabitants of the Mississippi delta were not only afraid of technical failure such as crevasses, but also of their own fellow men: levee guards were "men assigned to patrol the riverfront in order to protect levees from being cut" and were hired by property owners to prevent "unauthorized visits to the ramparts and preventing all visits at night" (ibid.: 112). Nevertheless, it was decided to cut the downstream levees.[14]

In stark contrast to the overall situation marked by strong social cohesion in the Netherlands, these examples illustrate that neither the citizens of New Orleans nor the city and its hinterlands understood themselves as a community of interests. Rather, the burdens associated with flooding were shifted from the powerful center to peripheral areas (within the city as well as between the city and its rural hinterlands).

Another interesting example, taken from the realms of public health and the installation of sewage and waterlines, dates back to the late nineteenth and early twentieth century (only 100 years before Hurricane Katrina hit the city).

> [...] at the close of the nineteenth century New Orleans residents still lived [...] on a giant "dung heap" inadequately drained by several open canals emitting effluvia to a common resource, the atmosphere. [...] The economic costs of this filth eventually inspired a civic response. When property values along the drainage canals depreciated due to the offensive odor, civic leaders claimed that proper drainage would restore them.
>
> Colten 2005: 56

Affluent neighborhoods were the first to be provided with water supply and sewage systems. In the low-lying areas, in contrast, sewage and feces accumulated in small pools. As a result of the high groundwater level, these did not seep away but moldered in the muggy subtropical climate. Thus, inhabitants of the low-lying areas were not only more likely to be affected by yellow fever epidemics or other health issues, but also "drinking a fluid likened to 'graveyard water' remained an overwhelming likelihood" (ibid.: 61). Only once the situation in the low-lying areas had deteriorated to such an extent that the unsanitary conditions frustrated economic growth of the whole city did the city council take action. "Sanitary improvements, even in areas occupied by blacks, would serve the entire city by reducing disease threats and thereby enhancing economic opportunities" (ibid.: 83). In the case of the sewage system, however, implementation was hampered by the fact that connection "to the sewerage system required a deposit of either $25 or $50 and thereby limited access to property owners with the resources to pay the fee" (ibid.: 90). This practice resulted not only in a piecemeal and suboptimal provision of public infrastructure and related goods such as public health, but was also economically inefficient in that positive effects of scale associated with network technologies could not be realized. Nevertheless, and true to the motto "you get what you pay for", this procedure satisfied the liberally biased world view and its fixation on market exchange as the tool of choice for coordination.

Taken together, what overall conclusions can be drawn from these observations and considerations? To give a short answer, and paraphrasing Schröder (2009: 32), the examples presented above elucidated how "hegemonic belief systems" influence the way in which social actors approach socio-ecological tasks and challenges. According to Giddens (1984), it can be claimed that culturally biased ways of life constitute crucial elements of practical consciousness and – for that reason – guide the way in which social actors "think about challenge, crisis and change, meaning certain options are automatically off the table and others are seen to be more legitimate" (Schröder 2009: 32). Culturally biased ways of thinking and feeling translate into social action (or inaction) which – at least in the case of New Orleans – becomes manifest in spatial and racial segregation, social inequality, and spatially unequal provision of public infrastructure and related goods and life chances. From an analytical perspective, such constellations can be described as eco-cultural mismatches. Once this mélange between physical infrastructure and corresponding social structures has come into being, it takes on a more and more persistent and recalcitrant character (Veblen 1990, Brette 2003): all the more so, as occurring mismatches are idealized and legitimated by appropriate ideational interpretive patterns such as the "up by your own bootstrap" philosophy.

Staying with the topic of mismatches, the example of New Orleans also elucidates – in contrast to the Netherlands – that the sheer existence of cooperative advantages does not automatically or deterministically result in cooperative behavior. Once hegemonic cultural preferences and related institutional and regulatory patterns have come into being, they tend to become relatively autonomous and to impose themselves on individual habits, physical nature, and new institutions so as to make them consistent with itself (Brette 2003) – no matter whether cooperative advantages are bargained away or not. In this context, and referring to the findings of Bednar and Page (2007) according to which hegemonic cultural and institutional structures are of higher overall societal efficiency even though resulting in case-specific mismatches, New Orleans and its deltaic wetlands seem to present one of these cases (together with, for example, the semiarid Great Plains, cf. Worster 1979). Thus, the example of New Orleans and its environs arguably illustrates that institutional patterns and cultural preferences that suffice for frontier societies are simply inappropriate in ecologically sensitive landscapes such as deltaic wetlands, for example, because cooperative advantages – as in the case of New Orleans – are not utilized on ideational grounds. This results not only in piecemeal solutions – that is, suboptimal adaptation to physical nature, for example, "backdoor flooding" due to different qualities or heights of levees – but also in economic inefficiencies. Owing to the fact that the provision of public infrastructure was often outsourced to private enterprises (at least in the past), economic inefficiencies commonly translated into insufficient maintenance work and – for that reason – into suboptimal functioning.

Against the background of these observations and considerations, the devastation of New Orleans by Hurricane Katrina and its humanitarian aftermath in 2005 seems to be the logical consequence of liberally biased eco-cultural path-dependencies. In other words, and comparably to the proverb of the "new wine in old wineskins", Hurricane Katrina can be seen as just another extreme weather event which hit the city and its well-established (but insufficient) liberally biased infra- and social structure. Of course, approximately 1,800 fatalities and thousands of displaced people represent heavy burdens, but from the perspective of eco-cultural adaptation, history simply repeated itself. For further reading on the humanitarian aftermath of Hurricane Katrina, the reader is referred to Michael Dyson (2007), Chester Hartman and Gregory Squires (2006), Kristin Bates and Richelle Swan (2007), Jenni Bergal (2007), Hillary Potter (2007), and David Brunsma et al. (2010).

Even though technologically driven improvements can be observed for the twentieth century, for example, with regard to sewage or drainage in low-lying areas, the relevant literature is in agreement on the fact that the topic of flood protection was of secondary importance before Katrina hit the city (Bergal 2007, Dyson 2007, Colten 2009). As elaborated by Colten (2009), hydrological interference with the delta and its sensitive ecosystem primarily had to serve economic interests: for example, the creation of either navigable canals (meaning that the meandering Mississippi river was cut through) or habitable space to gain a larger tax basis. Unintended negative side effects such as the loss of coastal wetlands – that is, loss of natural buffers to storms – or the artificial creation of "hurricane highways" via dead-straight shortcuts linking the Gulf of Mexico with the harbor of New Orleans were apparently not taken into account (Hiles 2007).

By 2005, this overall socio-structural situation had not changed significantly. Spatial and racial segregation and related socio-demographic clusters (for example, in terms of income, health, or age) could still be observed.[15] In other words, affluent households still lived on heightened grounds or in better-protected areas whereas lower-income households predominantly resided below sea level. Whereas affluent households still had the private means to use marketable coping strategies such as cars (to leave the city), flood insurance, and rebuilding of their homes, the coping capacities of poor households residing in low-lying areas were very limited: they could rely neither on marketable coping strategies nor on welfare arrangements (due to low decommodification). To give an example, despite calls for evacuation, the majority of the poor (and mostly black) residents living in low-lying areas were not able to leave the city simply because they did not have a car (Dyson 2007). Moreover, individual characteristics such as poor health or age negatively impinged on individual coping capacities. Deprived in such a way, no wonder the poor had to make use of individually applicable coping strategies such as looting or theft (for example, to procure food and medicine) or simply singing prayers while clinging to the roofs of their drowning homes and waiting for help.

Soon, however, they were criminalized for looting and – in doing so – violating private property rights.

To put it more pointedly: the government response to Hurricane Katrina came pretty late and was seen only when private property rights – that is, one of the most central constituents of liberally biased eco-cultural habitats – were publicly called into question by (so-called) looters:

> [S]hortly after Katrina struck, media reports began to converge on a series of images that were presented as typical behaviors being undertaken by disaster victims in New Orleans. Those images highlighted social breakdown, lawlessness, and violence. New Orleans was depicted as a "snakepit of lawlessness and anarchy," and victims fighting for survival in unimaginably traumatic circumstances were characterized as criminals and "marauding thugs."
>
> Tierney/Bevc 2010: 41

Thus, Iraq battle-tested troops "armed with loaded weapons to deal with socially constructed threats of 'urban insurgents' and charged with restoring order" (ibid.: 35) were sent to the city, authorized to shoot and kill (Dyson 2007). Just as ecological sustainability and flood protection did not play a major role in molding the deltaic wetlands but were subordinated to economic interests, it appears that humanitarian aid was secondary to reinforcing property rights.

From this perspective, and to cut a long story short, the devastation of New Orleans by Hurricane Katrina and its humanitarian aftermath can be described as just another "act" in the city's "eco-cultural drama", following a liberally biased "script" played out on a rather fragile ecological and social stage.

6.3 Synthesis and further research

Starting with the question of how different approaches to physical nature and society along the Rhine–Meuse–Scheldt and Mississippi deltas can be explained, this chapter has tried to elaborate on the premise that these differences are due to path-specific forms of institutional specialization and corresponding cultural preferences. In this pursuit, and by using the relevant literature about elective affinities between welfare and production regimes as well as related modes of regulating human interference with social and physical nature, a first synoptic picture describing the institutional and regulatory particularities of liberal and coordinated paths of eco-cultural adaptation was drawn in subchapter 6.1. Based on this "institutional inventory", the questions of how culturally biased modes of life along the Rhine–Meuse–Scheldt and Mississippi deltas initially originated and how they affect path-specific coping capacities with regard to social and ecological tasks and challenges occupied center stage in subchapter 6.2. Using the example of flood protection along the Rhine–Meuse–Scheldt delta, the first question was paradigmatically answered by illustrating how dilemmas of collective action could successfully be resolved and how well-adapted linkages between physical and social nature emerged over the course of time. Thereby, an eco-deterministic perspective was pursued. Using these considerations as a foil of comparison, the second question was elucidated by discussing the liberally inspired attempts at providing public infrastructure and related goods in New Orleans and its deltaic wetlands.

Trying to conflate the "institutional inventory" of subchapter 6.1 and the case studies of subchapter 6.2, it appears that the initial theses can be confirmed (at least for the time being): first, different approaches to physical nature and society along the deltas of both the Rhine–Meuse–Scheldt and Mississippi are rooted in different kinds of path-specific institutional specialization and corresponding cultural preferences. Second, the long-lasting struggles to provide public infrastructure in New Orleans and its deltaic environs and its devastation by Hurricane Katrina can be explained by fundamental structural mismatches between a liberally biased mode of life on the one side and the intricacies of deltaic wetlands in combination with the institutional and cultural prerequisites for public goods on the other. The differences between coordinated and liberal paths of adaptation (and related path-dependencies) become observable in the following way: the coordinated regime type emphasizes the principle of precaution and is specialized in the provision of public goods and related regulatory principles (especially long-term planning and centralized coordination combined with strong social cohesion). Liberally biased regimes, in contrast, primarily trust in the coordinative abilities of the market within an overall institutional environment characterized by somewhat laissez faire-like conditions (and the principle of aftercare).

In the case of the Netherlands, this finds expression in substantial entitlements with regard to publicly financed welfare benefits which also comprises the provision of various public goods by the state (for example, public security, health services, an educational system, or flood protection). Consequently, the overall level of social inequality is comparably low. As a result of similar regulatory approaches to physical nature and society, human interference with physical nature is shaped by comparable patterns. Just as these define social standards such as minimum income, age of retirement, or gender justice, they set up standards regarding contaminant inputs, fishing quota, size and number of polders, or energy consumption of refrigerators and light bulbs to maintain what is assumed to be environmentally sustainable.

The converse is the case for the liberal path of adaptation. Government intervention and state paternalism are unwanted and are perceived as illegitimate interference in private matters. On these ideational grounds, centralized coordination – for example, to provide public goods such as health insurance for everybody – is often rejected. Accordingly, the liberal habitat only provides the absolute minimum of state-funded welfare benefits. The degree of social inequality is comparably high. Anything that goes beyond this minimum standard – be it public health, public security, education, or unemployment benefits – has to be individually achieved, first and foremost via market exchange. To a large extent, goods such as public security are commodified, for example, in terms of guns, panic rooms, gated communities, or private security services. Those who cannot make use of market mechanisms to gain more than the minimum welfare publicly accessible must rely on kinship solidarity, charity movements, or simply have to pull themselves up by their own bootstraps (otherwise, failure is their own fault).

Thus, and in contrast to the overall situation in the Netherlands, social actors in the liberal habitat seem to neglect each other's welfare under the cloak of (alleged) equality of opportunity and justice of performance. In the same way as social actors constitute their interrelationships in terms of competitive behavior and market coordination, the social constitution of physical nature and its valorization also obeys the competitive logic of commodification and the short-term maximization of individual utility (as opposed to the maximization of collective utility and long-term planning in the Netherlands). Comparable to labor skills, physical nature is perceived as a resource (or factor of production in the broadest sense) whose value is (directly or indirectly) determined via market mechanisms. Analogous to the principle of "hire and fire", which allows companies to employ or dismiss employees according to short-term requirements, it appears that human interference with nature and the exploitation of natural resources is also oriented on considerations of short-term utility maximization. Believing in technological progress and substitutability, notions of long-term sustainability or limited carrying capacity are put aside (Solow 1974, Stiglitz 1979, Daly 1997). In this sense, and in contrast to the Netherlands, inhabitants of the liberal habitat show solidarity neither with their fellow men nor with their surroundings in general. However, there are specific exceptions, for example, in the case of national parks, a fact which – on closer inspection – turns out to primarily serve the maintenance of the self-image of being a frontier society and hedging the "Wild West" mythology (Isenberg 1997).

In terms of resilience and vulnerability (Adger/Kelly 1999), these considerations and observations can be condensed in the following way: in contrast to the Dutch, the inhabitants of New Orleans and its environs are vulnerable to flooding in a twofold way. First, and as a result of the inadequate institutional embeddedness of hydraulic works and resulting piecemeal measures and economic inefficiency, the general level of preparedness is suboptimal. Thus, flooding is more likely to happen. Second, as a result of low levels of decommodification and the resulting social inequalities, individual capacities to cope with and recover from flooding are unequally distributed. This was demonstrated quite plainly by Hurricane Katrina. Yet, and due to strong path-dependencies, no happy ending to New Orleans' "eco-cultural drama" or significant changes in its liberally biased "script" are in sight. It is probably just a matter of time before history will repeat itself once again. However, and because the most severely affected quarters such as the Lower 9th Ward (located in the eastern outskirts) have more or less been abandoned since Hurricane Katrina, the hope remains that upcoming extreme weather events do not turn into social disasters.[16]

On the basis of these findings, what theoretical conclusions can be drawn and which questions for further research can be envisioned? In general, the present chapter has shown that the approach of eco-cultural adaptation is a fruitful theory. After Chapter 5 delivered robust evidence of its quantitative operationalizability and predictive capacity, the present chapter has demonstrated that it also yields substantial explanatory power when qualitatively and historically inspired research questions occupy center stage. Thereby, the analytical advantage of historical-reconstructive approaches can be seen in the fact that these allow for a deeper understanding of observable eco-cultural fabrics by comprehending their emergence in retrospect (for example, Sieferle et al. 2006 or Acemoglu and Robinson 2012). To give an example, and in contrast to Chapter 5, it is possible, for example, to account for changes in time, to scrutinize the Janus-faced character of eco-cultural habitats (in terms of specialization and related vulnerabilities), or to understand how and why environmental change such as land subsidence results in technological, institutional, and cultural adjustments (or vice versa). The present chapter has produced empirical evidence confirming, for example, the theoretically assumed tension between functional necessities and cultural preferences or the similar constitution of physical nature and society.

In this context, the compatibility between different regulatory approaches regarding social welfare, production process, and ecological issues (Esping-Andersen 1990, Scruggs 1999, Hall/Soskice 2001, Meadowcroft 2005, Duit 2008) and the approach of eco-cultural adaptation is particularly interesting. This not only corroborates the theoretical ideas of path-specific institutional specialization and the similar constitution of physical and social nature, but also presents an interesting topic of further research, namely to expand the well-established typologies of corresponding welfare and production regimes by a matching and profound typology of eco-regimes. In this context, the approach of eco-cultural adaptation as developed here might serve as an overall frame of reference, allowing the embedding of existing approaches to welfare and production regimes and corresponding regulatory approaches to ecological issues in a greater overall context. From this perspective, it could be argued that the well-established typologies of Esping-Andersen (1990) and Hall and Soskice (2001) problematize special cases within a much larger overall context, namely the interplay between physical nature and society.

Staying with the topic of further research, three further fields appear to be of interest, especially with regard to theory building and consolidation. First, it would be rewarding to learn more about how public infrastructure is provided in other cities in North America. In other words, is the situation in New Orleans representative of most cities within the liberal North American habitat? Or does it constitute an exception? If it constitutes an exception, how do other cities succeed in providing public infrastructure? In this context, and with regard to the idea of fractalization as discussed in the Introduction, it could be argued that cities represent "coordinated islands" within a comparably uncoordinated eco-cultural habitat.

Second, it would be worthwhile to examine the provision and functioning of technologies other than levees in North America and the Netherlands. In this context, it can be assumed that the dissemination of unconnected technologies such as wind turbines should proceed more rapidly and with fewer complications in liberal than in coordinated adaptive paths. Indeed, tentative hints from Texas can be found in this context (Langniss/Wiser 2003, Menz/Vachon 2006, Wiser/

Barbose 2008). Although Texas rejects any environmental political guidelines from Washington and has a strong reputation for not caring much about environmental niceties, it is about to become one of the world leaders with regard to regenerative energies. According to their liberal bias, the exploitation of wind – rather than oil – is just another way of making money and of securing their liberal way of life (in this case: to be independent of foreign oil and gas). The converse is the case in European states such as Germany. In densely populated areas, erecting wind parks is an intricate endeavor as a result of strict admission procedures and high negotiation costs among investors, environmentalists, residents, and municipalities.

Third and finally, the approach of eco-cultural adaptation should be applied to countries other than highly developed Western countries and related forms of adaptation. Here again, the work of Bevan (2004), Wood (2004), Wood and Gough (2006), and Gough (2007) – dealing with well-being in developing countries and respective (in-)security regimes – would present a suitable point of entrance. In order to guarantee a minimum of comparability, another delta such as the Ganges–Brahmaputra in Bangladesh would be well suited for an analysis of how the eco-cultural fabric is assembled under the conditions of comparably low degrees of labor division and specialization, related low-tech solutions to flooding (for example, floating gardens or seasonal migration patterns), and solidarity restricted to kin or clan (Zaman 1999, Islam/Atkins 2007).

Notes

1 In the following, the discussion is limited to liberal and coordinated production regimes as described by the Varieties of Capitalism approach, that is: First World countries. Generally, and following Marx (1932a, 1932b), according to whom labor and production processes are the pivotal point for securing a livelihood, it is claimed that the following considerations should also be transferable to so-called developing and emerging countries or subsistence economies. In this context, the work of Bevan (2004), Wood (2004), and Gough (2007) dealing with welfare and (in-)security regimes in developing countries would present a suitable starting point. Here, and as elaborated by North (1990: 34 ff.) and in Chapter 2, exchange relations are more likely to be coordinated by clientelism or kinship (as opposed to third-party enforcement as is the case in liberal or coordinated welfare and production regimes). From this perspective, mafia-like structures coordinating the provision of specific goods and services can be seen as functional equivalents to the role taken by the Leviathan in liberal and coordinated welfare and production regimes.

2 Thereby, the "relational" firm is located at the center of the whole approach. In contrast to pure microeconomics, which primarily focuses on the individual behavior of households and firms, the Varieties of Capitalism approach rather addresses questions located at the junction of micro- and macro- as well as institutional economics. It is argued that the development of goods and services, as well as their production and distribution, crucially depends on "the quality of the relationships the firm is able to establish, both internally with its own employees, and externally with a range of other actors that include suppliers, clients, collaborators, stakeholders, trade unions, business associations, and governments" (Hall/Soskice 2001: 6). As indicated by the term "relational" firm, in turn, it is assumed that the firm's ability to coordinate and maintain sufficient relationships with internal and external actors largely depends on the

institutional environment and the regulatory regime in which it is embedded. However, entertaining exchange relations and coordinating cooperation is an intricate and costly endeavor (cf. Chapter 2): it is necessary not only to negotiate terms of trade, but also to monitor their compliance and to enforce them in case of violation (North 1990, Becker/Murphy 1992). Against the background of these considerations, specific forms of cooperation can be facilitated by regulatory regimes and related institutions, for example, by providing third-party enforcement (North 1990).

3 As a rule of thumb, the following applies: the higher the degree of decommodification, the lower the dependency on gainful employment (and vice versa). To give some examples, the degree of decommodification becomes observable in the specific configuration of minimum wages, benefit payments, dismissal protection, sickness benefits, pension schemes, or the properties of the health and educational system.

4 As argued by Dryzek (in Gough et al. 2008), this is also corroborated by the (contentious) sustainability index (www.yale.edu/esi). Here, "Finland comes first, with other Northern European states in close attendance [...]. These are also the countries with the most highly developed welfare states. Conversely, the more liberal, market-oriented countries of the Anglo-American world in particular have both lower social policy effort and weaker environmental policy performance" (ibid.: 334).

5 Floating gardens are "made of decomposing heaps of water hyacinth, with an upper surface layer of ash, coconut fibre and, occasionally, soil" (Islam/Atkins 2007) and can be moved by boats from one site in the delta to another.

6 As elaborated by TeBrake (2002), subsidence "occurred, first of all, because lowering the water table within a mass of peat reduced its volume, resulting in compaction of the plant material. [...] The immediate problem caused by subsidence was reduced gradient, which made it more difficult to achieve adequate drainage" (ibid.: 483 ff.).

7 To give an example, the St. Elisabeth's Day flood in 1421 "inundated 500 square kilometers and probably killed ten thousand people" (Reuss 2002: 467).

8 "The regional water authorities, in short, acquired considerable legislative, judicial, and executive powers over regional water management issues, while the village councils kept the responsibility for local hydraulic structures" (Kaijser 2002: 528).

9 Today, this is also reflected in the 'Room for the River' approach, which can be described as a new water management doctrine emphasizing the importance of retention areas for flood protection. As described by Roth and Winnubst (2009), the Dutch do not hesitate to resettle or even disappropriate those living in these areas to create new spaces for flood retention. The Oosterschelde storm surge barrier would be another example. Here, movable floodgates – four kilometers long in total – guarantee appropriate flood protection in case of high waters. Under normal meteorological circumstances the floodgates stay open and – in contrast to a closed dam – allow water exchange between the sea and the Oosterschelde, guaranteeing the preservation of the brackish environment and marine life behind the barrier (van de Ven 2004).

10 For visualization, the reader is referred to www.datacenterresearch.org (formerly the Greater New Orleans Community Data Center). Here, maps illustrating New Orleans' topography, the geographies of poverty, as well as ethnic spatial segregation are available.

11 From an analytical perspective, this means that "public" infrastructure here should rather be treated as club goods than public goods.

12 Since 1917, but especially since Hurricane Betsy in 1965, the US Army Corps of Engineers has been involved in flood protection in New Orleans and its environs. The simple fact that flood protection in New Orleans (and elsewhere in the United States) rests on the Army Corps of Engineers could be read as another hint indicating strong institutional path-dependencies. Apparently, there are no other governmental structures allowing for centralized coordination to provide public goods. Cognitively, it seems to make no difference whether US citizens and their marketplaces around the world are defended against either terror or undesirable "malfunctions" of physical nature.

13 In contrast to the Netherlands, only those living along the riverbanks had to bear the investments and responsibilities for flood protection. Those living in the hinterlands and profiting from the investments of their neighbors were not charged (at least in those times).

14 "The governor's decision to cut the levees downriver from the city, rather than upstream where the levee was already weak and in need of costly shoring, resulted from businessmen's reminders that the lands upriver were more developed, and reimbursement for damages there would be far more costly to the city. No doubt, Simpson [who was the Governor in those days, author's note] had also weighed the political ramifications of inundating the lands of wealthy and influential citizens in the sugar-producing parishes upriver from New Orleans" (Gomez 2000: 114).

15 www.gnocdc.org/prekatrinasite.html

16 In this context, emigration (exit) can be seen as a low-tech coping strategy for those who are unable to voice their right to better living conditions (Hirschman 1970).

Bibliography

Acemoglu, D., & Robinson, J. (2012). *Why nations fail. The origins of power, prosperity and poverty.* New York: Crown Publishing.

Adger, W. N., & Kelly, P. M. (1999). Social vulnerability to climate change and the architecture of entitlements. *Mitigation and Adaptation Strategies for Global Change,* 4(3/4), 253–266.

Bates, K. A., & Swan, R. S. (Eds.) (2007). *Through the eye of Katrina: Social justice in the United States.* Durham: Carolina Academic Press.

Becker, G. S., & Murphy, K. M. (1992). The division of labor, coordination costs, and knowledge. *The Quarterly Journal of Economics,* 107(4), 1137–1160.

Bednar, J., & Page, S. (2007). Can Game(s) Theory explain culture? The emergence of cultural behavior within multiple games. *Rationality and Society,* 19(1), 65–97.

Bergal, J. (Ed.) (2007). *City adrift: New Orleans before and after Katrina.* Baton Rouge: Louisiana State Univ. Press.

Bevan, P. (2004). Conceptualising in/security regimes. In I. Gough (Ed.), *Insecurity and welfare regimes in Asia, Africa, and Latin America. Social policy in development contexts* (1st ed., pp. 88–118). Cambridge: Cambridge Univ. Press.

Bijker, W. E. (2002). The Oosterschelde storm surge barrier: A test case for Dutch water technology, management, and politics. *Technology and Culture,* 43(3), 569–584.

Bohle, H. G., Downing, T. E., & Watts, M. J. (1994). Climate change and social vulnerability: Toward a sociology and geography of food insecurity. *Global Environmental Change,* 4(1), 37–48.

Bowles, S., Edwards, R., & Roosevelt, F. (2005). *Understanding capitalism: Competition, command, and change* (3rd ed.). New York: Oxford Univ. Press.

Brette, O. (2003). Thorstein Veblen's theory of institutional change: beyond technological determinism. *The European Journal of the History of Economic Thought,* 10(3), 455–477.

Brunsma, D. L., Overfelt, D., & Picou, J. S. (Eds.) (2010). *The sociology of Katrina: Perspectives on a modern catastrophe* (2nd ed.). Lanham: Rowman & Littlefield Publishers.

Cole, A. P., Adams-Fuller, T., Cole O. J., Kruglanski, A., & Glymph, A. (2007). Making sense of a hurricane: Social identity, and attribution of race-related differences in Katrina disaster response. In H. Potter (Ed.). *Racing the storm. Racial implications and lessons learned from Hurricane Katrina* (pp. 3–32). Lanham: Lexington Books.

Colten, C. E. (Ed.) (2000). *Transforming New Orleans and its environs: Centuries of change.* Pittsburgh: Univ. of Pittsburgh Press.

(2005). *An unnatural metropolis: Wresting New Orleans from nature.* Baton Rouge: Louisiana State Univ. Press.

(2009). *Perilous place, powerful storms: Hurricane protection in coastal Louisiana.* Jackson: Univ. Press of Mississippi.

Daly, H. (1997). Georgescu-Rogen versus Solow/Stiglitz. *Ecological Economics*, 22, 261–266.

Davis, D. W. (2000). Historical perspective on crevasses, levees, and the Mississippi River. In C. E. Colten (Ed.), *Transforming New Orleans and its environs. Centuries of change* (pp. 84–106). Pittsburgh: Univ. of Pittsburgh Press.

Duit, A. (2008). The Ecological State: Cross-National Patterns of Environmental Governance Regimes, *EPIGOV Paper No. 39*. Berlin: Ecologic-Institute for International and European Environmental Policy.

Dyson, M. (2007). *Come hell or high water. Hurricane Katrina and the color of disaster.* New York: Basic Civitas.

Elkins, S. M., & MacKitrick, E. (1993). *The age of federalism.* New York: Oxford Univ. Press.

Ellickson, R. (1991). *Order without law: How neighbors settle disputes.* Cambridge, MA: Harvard Univ. Press.

(1993). Property in Land. *Faculty Scholarship Series*, Paper 411. http://digitalcommons. law.yale.edu/fss_papers/411

Esping-Andersen, G. (1990). *The three worlds of welfare capitalism.* Cambridge: Polity.

Estevez-Abe, M., Iversen, T., & Soskice, D. (2001). Social protection and the formation of skills: A reinterpretation of the welfare state. In P. A. Hall & D. Soskice (Eds.), *Varieties of capitalism. The institutional foundations of comparative advantage* (pp. 145–183). Oxford: Oxford Univ. Press.

Foner, E. (1995). *Free soil, free labor, free men: the ideology of the Republican Party before the Civil War.* Oxford: Oxford Univ. Press.

Fussell, E., Sastry, N., & VanLandingham, M. (2010). Race, socioeconomic status, and return migration to New Orleans after Hurricane Katrina. *Population and Environment*, 31(1–3), 20–42.

Giddens, A. (1984). *The constitution of society: Introduction of the theory of structuration.* Berkeley: Univ. of California Press.

Gomez, G. M. (2000). Perspective, power, and priorities: New Orleans and the river flood of 1927. In C. E. Colten (Ed.), *Transforming New Orleans and its environs. Centuries of change* (pp. 109–120). Pittsburgh: Univ. of Pittsburgh Press.

Gough, I. (Ed.) (2007). *Wellbeing in developing countries: From theory to research* (1st ed.). Cambridge: Cambridge Univ. Press.

et al. (2008). JESP symposium: Climate change and social policy. *Journal of European Social Policy*, 18(4), 325–344.

Gough, I., & Meadowcroft, J. (2011). Decarbonizing the welfare state. In J. S. Dryzek, R. B. Norgaard, & D. Schlosberg (Eds.), *The Oxford handbook of climate change and society* (pp. 490–503). Oxford: Oxford Univ. Press.

Hall, P. A., & Soskice, D. (Eds.) (2001). *Varieties of capitalism: The institutional foundations of comparative advantage.* Oxford: Oxford Univ. Press.

Hartman, C. W., & Squires, G. D. (Eds.) (2006). *There is no such thing as a natural disaster: Race, class, and Hurricane Katrina.* New York: Routledge.

Henrich, J., Boys, R., Bowles, S., Camerer, C., Fehr, E., & Gintis, H. (Ed.) (2004). *Foundations of Human Sociality: economic experiments and ethnographic evidence from fifteen small-scale societies.* Oxford: Oxford Univ. Press.

Henrich, N., & Henrich, J. (2007). *Why humans cooperate: A cultural and evolutionary explanation.* Oxford: Oxford Univ. Press.

Hiles, S. S. (2007). The Environment. In J.E.A. Bergal (Ed.), *City adrift. New Orleans before and after Katrina* (pp. 7–19). Baton Rouge: Louisiana State Univ. Press.

Hirschman, A. O. (1970). *Exit, voice, and loyalty: Responses to decline in firms, organizations, and states.* Cambridge: Harvard Univ. Press.

Hudson, P. F., Middelkoop, H., & Stouthamer, E. (2008). Flood management along the Lower Mississippi and Rhine Rivers (The Netherlands) and the continuum of geomorphic adjustment: The 39th Annual Binghamton Geomorphology Symposium: Fluvial Deposits and Environmental History: Geoarchaeology, Paleohydrology, and Adjustment to Environmental Change. *Geomorphology*, 101(1–2), 209–236.

Isenberg, A. C. (1997). The returns of the bison: Nostalgia, profit, and preservation. *Environmental History*, 2(2), 179–196.

Islam, T., & Atkins, P. (2007). Indigenous floating cultivation: a sustainable agricultural practice in the wetlands of Bangladesh. *Development in Practice*, 17(1), 130–136.

Iversen, T., & Soskice, D. (2001). An Asset Theory of Social Policy Preferences. *The American Political Science Review*, 95(4), 875–893.

Kaijser, A. (2002). System building from below: Institutional change in Dutch water control systems. *Technology and Culture*, 43(3), 521–548.

Lamb, H. (1991). *Historic storms of the North Sea, British Isles and Northwest Europe.* Cambridge: Cambridge Univ. Press.

Langniss, O. & Wiser, R. (2003). The Renewables Portfolio Standard in Texas: An early assessment. *Environmental Policy* 31(6), 527–535.

Lintsen, H. (2002). Two centuries of central water management in the Netherlands. *Technology and Culture*, 43(3), 549–568.

Marx, K. (1932a). *Das Kapital: Kritik der politischen Ökonomie* (2nd ed.). Berlin: Kiepenheuer. (1932b). *Die deutsche Ideologie: Kritik d. neuesten deutschen Philosophie in ihren Repräsentanten, Feuerbach, B. Bauer u. Stirner, u. d. deutschen Sozialismus in seinen verschiedenen Propheten 1845–1846* (1st ed.). Vienna: Verl. f. Literatur u. Politik.

Maslow, A. H. (1966). *The psychology of science; A reconnaissance.* New York: Harper & Row.

McKean, M. A. (1996). Common-property regimes as a solution to problems of scale and linkage. In S. S. Hanna (Ed.), *Rights to nature. Ecological, economic, cultural, and political principles of institutions for the environment* (pp. 223–244). Washington, D.C: Island Press.

Meadowcroft, J. (2005). From welfare state to ecostate? In J. E. Barry (Ed.), *The state and the global ecological crisis* (pp. 3–23). Cambridge: MIT Press.

Menz, F. C., & Vachon, S. (2006). The effectiveness of different policy regimes for promoting wind power: Experiences from the states. *Energy Policy*, 34(14), 1786–1796.

North, D. C. (1990). *Institutions, institutional change, and economic performance.* Cambridge: Cambridge Univ. Press.

Olson, M. (1965). *The logic of collective action; public goods and the theory of groups.* Cambridge: Harvard Univ. Press.

Poteete, A. R., Janssen, M. A., & Ostrom, E. (2010). *Working together: Collective action, the commons, and multiple methods in practice.* Princeton: Princeton Univ. Press.

Potter, H. (Ed.) (2007). *Racing the storm: Racial implications and lessons learned from Hurricane Katrina.* Lanham: Lexington Books.

Reuss, M. (2002). Learning from the Dutch: Technology, management, and water resources Development. *Technology and Culture*, 43(3), 465–472.

Roth, D., & Winnubst, M. (2009). Reconstructing the polder: negotiating property rights and 'blue' functions for land. *International Journal of Agricultural Resources, Governance and Ecology*, 8(1), 37–56.

Schreuder, Y. (2001). The polder model in Dutch economic and environmental planning. *Bulletin of Science, Technology & Society*, 21(4), 237–245.

Schröder, M. (2009). Integrating welfare and production typologies: How refinements of the varieties of capitalism approach call for a combination of welfare typologies. *Journal of Social Policy*, 38(1), 19–43.

Scruggs, L. A. (1999). Institutions and environmental performance in seventeen Western democracies. *British Journal of Political Science*, 29(1), 1–31.

Sieferle, R. P. (2011). Cultural evolution and social metabolism. *Geografiska Annaler: Series B, Human Geography*, 93(4), 315–324.

et al. (Ed.) (2006). *Das Ende der Fläche: Zum gesellschaftlichen Stoffwechsel der Industrialisierung*. Cologne: Böhlau.

Solow, R. M. (1974). The economics of resources or the resources of economics. *The American Economic Review*, 64(2), 1–14.

Stavin, R. N (2011). The problem of the commons: Still unsettled after 100 years. *The American Economic Review*, 101, 81–108.

Stiglitz, J. E. (1979). A neoclassical analysis of the economics of natural resources. In V. K. Smith (Ed.), *Scarcity and growth reconsidered* (pp. 36–66). Baltimore: Johns Hopkins Univ. Press.

Syvitski, J. P., Kettner, A. J., Overeem, I., Hutton, E.W., Hannon, M., Brakenridge, G. R., Day, J., Vörösmarty, C., Saito, Y., Giosan, L., & Nicholls, R. J. (2009). Sinking deltas due to human activities. *Nature Geoscience*, 2(10), 681–686.

TeBrake, W. H. (1995). *Medieval frontier: Culture and ecology in Rijnland*. College Station: Texas A&M Univ. Press.

(2002). Taming the waterwolf: Hydraulic engineering and water management in the Netherlands during the Middle Ages. *Technology and Culture*, 43(3), 475–499.

Tierney, K., & Bevc, C. (2010). Disaster as war: Militarism and the social construction of disaster in New Orleans. In D. L. Brunsma, D. Overfelt, & J. S. Picou (Eds.), *The sociology of Katrina. Perspectives on a modern catastrophe* (2nd ed., pp. 35–49). Lanham: Rowman & Littlefield Publishers.

Turner, F. J. (1894). *The significance of the frontier in American history*. Marlborough: Adam Matthew Digital.

US Army Corps of Engineers (2014). www.usace.army.mil/About/MissionandVision.aspx (Jan. 6, 2014).

van Dam, P. J. (2002). Ecological challenges, technological innovations: The modernization of sluice building in Holland, 1300–1600. *Technology and Culture*, 43(3), 500–520.

van de Ven, G. P. (2004). *Man-made lowlands; history of water management and land reclamation in the Netherlands* (4th ed.). Utrecht: Uitg. Matrijs.

Veblen, T. (1990). *The place of science in modern civilization and other essays*. New Brunswick: Transaction Publishers [1919].

Watts, M. J., & Bohle, H. G. (1993). The space of vulnerability: the causal structure of hunger and famine. *Progress in Human Geography*, 17(1), 43–67.

Wiser, R. & Barbose, G. (2008). Renewables Portfolio Standards in the United States. A Status Report with Data through 2007, Lawrence Berkeley National Laboratory, http://emp.lbl.gov/sites/all/files/REPORT%20lbnl-154e-revised.pdf (Jan. 2, 2014).

Wittfogel, K. A. (1957). *Oriental despotism; a comparative study of total power.* New Haven: Yale Univ. Press.

Wood, G. (2004). Informal security regimes: the strength of relationships. In I. Gough (Ed.), *Insecurity and welfare regimes in Asia, Africa, and Latin America. Social policy in development contexts* (1st ed., pp. 49–87). Cambridge: Cambridge Univ. Press.

Wood, G., & Gough, I. (2006). A comparative welfare regime approach to global social policy. *World Development*, 34(10), 1696–1712.

Worster, D. (1979). *Dust Bowl.* Oxford: Oxford Univ. Press.

Zaman, M. Q. (1999). Vulnerability, disaster, and survival in Bangladesh: Three case studies. In A. Oliver-Smith, & S. M. Hoffman (Eds.), *The angry earth. Disaster in anthropological perspective* (pp. 192–212). New York: Routledge.

7 Résumé

In describing and explaining the relationship between human society and its physical environs from a coevolutionary perspective, the considerations presented here deviate from classic approaches of environmental sociology which – except for some rare examples – mostly address this subject from the unidirectional perspectives of materialism or constructivism. Here, the coevolutionary perspective provides a theoretical frame of reference which allows the integration of theoretical considerations from ecological economics into environmental sociology. Likewise, the coevolutionary theoretical frame of reference lends itself to enriching current approaches from ecological economics by an informed notion of society. Taken together, the approach of eco-cultural adaptation as developed here may allow new insights into the intricate relationship between physical nature and society, provide stronger explanatory power and predictive capacity, and represent a first step toward a critical theory of industrial societies which tries to overcome the separation of sociological and ecological thinking.

The approach of eco-cultural coevolution is illustrated by the theoretical figure of eco-cultural fabrics. By spelling out this theoretical frame of reference, particular attention has been paid to the following questions: how do cultural preferences emerge under the influence of different physical environments favoring either solitary action or cooperative behavior? And how do these cultural preferences affect the adaptive capacities of eco-cultural habitats and their inhabitants in the long run?

Work on the approach of eco-cultural coevolution promised the establishment of an integrative theoretical linkage and frame of reference facilitating the connection of theoretical currents from various scientific fields, especially from environmental sociology and psychology, as well as from evolutionary, environmental, institutional, behavioral, and cultural economics. However, this quality alone does not necessarily translate into desirable characteristics such as operationalizability, explanatory power, or generalizability and predictive capacity.

To test the empirical sensibility, the approach of eco-cultural coevolution was examined for the following questions: would it be possible to operationalize the concept of eco-cultural adaptation on different spatial scales and to apply it to both quantitative and qualitative research questions? Would the concept yield the theoretically expected results regarding habitat-specific correspondences among physical nature, technologies, institutions, and cultural value orientations? Would the selected examples actually show the theoretically expected spatial self-similarities?

These questions were first addressed from a quantitative perspective (Chapter 5). By intertwining the approach of eco-cultural adaptation with the concept of social metabolism as described by Sieferle et al. (2006), rural and urban modes of life were conceptualized as specific paths of eco-cultural adaptation. The leading assumption was that rural and urban paths of adaptation are characterized by distinct matches among physical nature, technologies, different modes of social reproduction, and cultural preferences and that these habitat-specific patterns can be observed on different analytical scales – an assumption which could be broadly corroborated by the empirical data.

To complete the picture, the conceptual questions sketched out above were approached from a qualitative and historic-reconstructive perspective on the basis of a paradigmatic comparison of two delta regions and the respective eco-cultural paths of adaptation: the Rhine–Meuse–Scheldt and Mississippi deltas (Chapter 6). In doing so, the crucial questions of how eco-cultural paths of adaptation take shape in the course of time and – once established – affect the coping capacities of their inhabitants occupied center stage. It was shown that the constitution of physical nature and society evolves along similar principles within one and the same eco-cultural path of adaptation. It was concluded that ecological and social tasks and challenges are approached in similar ways within one and the same habitat of eco-cultural adaptation, even though this might result in suboptimal outcomes in some cases.

All in all, and despite some ambivalent empirical results, one may infer that the concept of eco-cultural adaptation has the potential to come up to its own aspirations: it presents an analytical frame of reference capable of describing and explaining the intricate relationship between physical nature and society both from a materialist and social-constructivist perspective. It proves to be operationalizable – both in quantitative and qualitative terms and on different analytical scales – and yields the theoretically expected results, at least in most cases.

However, more research has to be done: first, to better understand and deepen the perspective developed here; and second, to assess the theoretical range of the approach of eco-cultural adaptation more systematically. Empirically, this could be achieved by confronting the approach of eco-cultural adaptation with such conditions as may be found in, for example, non-Western countries or in slum and subsistence economies. However, assessing the theoretical range of the concept developed here is not just an empirical question but also a theoretical one – especially with regard to its predictive capacity. Is it admissible to describe the concept of eco-cultural adaptation as a universally valid theoretical frame of reference? Or would it be more appropriate to describe it as a heuristic merely applicable to a limited number of questions and cases?

From the perspective of coevolutionary ecological economics, as, for example, represented by Norgaard (1994, 1997), the answer to these questions is quite simple: in the evolutionary paradigm, explanations make recourse to the fundamental mechanisms of variation, inheritance, and selection. Because these mechanisms occur incidentally, evolutionary processes are principally not predictable but represent historically contingent events.

> And depending on genetic mutations, value shifts, technological changes and social innovations that arise randomly, the evolutionary path is reset for a period until another change occurs. Thus the coevolutionary perspective explains why options are disturbingly limited in the short run: culture has determined environment and environment has determined culture. At each point in time there is a near gridlock of coevolved knowledge, values, technologies, social organization and natural environment. Yet over the long run we approach the equally disturbing situation of nothing determining anything, that all will change in unpredictable ways.
>
> Norgaard 1997: 163

Eco-cultural path-dependencies (or gridlocks) that might result from historically contingent events are an exception – providing the respective path proves to be stable, it is possible to make predictions, for example, about the compatibility of technological innovations and predominant cultural preferences. To forecast the coevolutionary outcome of emerging eco-cultural habitats as well as their operative logics, however, is beyond the capacity of the evolutionary paradigm.

By introducing the principles of labor division, economies of scale, and allometry (cf. Chapter 2), the concept of eco-cultural adaptation as developed here introduces an alternative approach to explain and predict the emergence of eco-cultural habitats and their operative logics. Similar to Newtonian mechanics, these principles constitute historically invariant quasi natural laws. They are valid beyond historical accidents and social constructivism and – for that reason – should allow for universally valid predictions. Consequently, it was argued that strong cultural preferences for cooperative action emerge when cooperative behavior is rewarded by physical nature and when the benefits of cooperation outweigh its costs. Strong cultural preferences for solitary action, on the contrary, develop when the costs of cooperation outweigh its advantages. Thereby, the occurrence of cooperative advantages (or disadvantages) is mediated by the given technological opportunities.

Thus, two different epistemes pervade the previous chapters. While evolutionary approaches usually have to make recourse to historically contingent events and path-dependent processes, the concept developed here also suggests taking universally valid constituents into account. Perhaps this is the actual innovation of this book. It is not clear yet, however, in what way these epistemes are related to each other and which makes which contribution to the coming into existence of eco-cultural habitats and their operative logics. Here, clearly more research and modeling has to be done, especially with regard to the following questions: how long are eco-cultural paths of adaptation stable? On what does their stability primarily depend? How relevant are historically contingent events and the effects of path-dependence, and what are the implications of historically invariant forces such as the laws of allometry? Is it possible to define balances and tipping points?

It is clearly beyond the scope of this final chapter to answer these questions in detail. Nevertheless, some thoughts about the relationship between evolutionary and quasi mechanical explanations should be put forward to stimulate further research. After all, it can be derived from the case studies of Chapters 5 and 6 that both epistemes are relevant for describing and explaining eco-cultural fabrics, with Chapter 5 concentrating on the effects of economies of scale and allometric rules and Chapter 6 examining the coming into existence of path-dependencies and related effects of cultural self-reinforcement. So, how could the integration of these theoretical perspectives be further developed?

To begin with, it could be argued that physical nature provides "abiotic attractors" or "condensation cores" on which some evolutionary events in terms of eco-cultural coevolution will dock with a higher probability than others. Take mountain areas as an example: why is it that extensive grazing can be usually observed in the higher regions whereas valleys are mostly characterized by intensive (irrigated) farming? Of course, valley locations do not directly determine the emergence of intensive farming. It could be argued, however, that they make the emergence of intensive farming more probable by rewarding intensive farming with higher calorie rates than extensive grazing.

On the other hand, it could be argued that eco-cultural paths of adaption do not depend on such attractors or condensation cores at all. Take cities, for example. Physical density and related advantages such as short distances or social contagion (Bettencourt et al. 2007) do not exist by nature and do not necessarily depend on abiotic attractors or such like. They may arise in any places where – for whatever reason – initial agglomeration took place, for example, at the junction of trade routes. These initial advantages can feed upon themselves and trigger a self-reinforcing process of increasing agglomeration and related advantages (Krugman 2009). Thus, paths of eco-cultural adaptation do not necessarily depend on mechanical fixations a priori. Society itself produces such condensation cores for agglomerations and steers their development.

Nevertheless, and staying with the example of agglomerative advantages and disadvantages, it can be argued that the historically invariant principles of labor division, economies of scale, and allometry begin to affect the course of agglomerations once a self-reinforcing process has developed. Probably, these principles (and their evolutionary advantages) are the reason why urban agglomerations exist at all. As argued in Chapter 2, however, the cooperative advantages that can be derived from these principles cannot be limitlessly increased, but are subject to specific capacity constraints. Thus, increasing concentration and agglomeration will eventually turn into cooperative disadvantages and – rather than short distances – we will experience congestion and the positive effects of high innovation rates will be counteracted by higher crime and infection rates (Bettencourt et al. 2007). In other words: even though the emergence of agglomerations might be subject to historically contingent events, their actual shape crucially depends on the historically invariant principles of labor division, economies of scale, and allometry. To date, however, it remains unclear when and why the advantages of agglomeration turn into disadvantages. Do they increase in the same manner? Is it possible to define tipping points? And how far can agglomerative disadvantages be mitigated by technological means?

Hopefully, these considerations will stimulate further research regarding the relationship of the two epistemes described above and result in a deeper understanding of the society–nature nexus. Apart from these academic considerations, it appears that the concept of eco-cultural adaptation offers considerable practical and political relevance: it could be used to make strong and valid statements about the compatibility among physical nature, technologies, institutions, and cultural value orientations. As the example of the German energy transition (but also many other technological projects) impressively illustrates, deeper knowledge about local constellations among physical nature, available technologies, predominant institutions, as well as related cultural value orientations is of crucial importance for its successful implementation.

Further, and as the comparison of rural and urban paths of adaptation has shown (Chapter 5), it could be concluded that each eco-cultural habitat has the potential to become more ecological according to its own manner. To give some examples, urban areas entail strong ecological potentials as a result of their higher compactness (for example, short distances and compact buildings with favorable surface-to-volume ratios); rural areas, in turn, are characterized by comparatively larger households and related saving potentials due to economies of scale within the household, for example, in terms of reduced living space or heating energy per capita. Accordingly, and in contrast to "one fits all" solutions, environmental regulation should be addressed in a more habitat-specific way to account for the local and historically grown entanglements among cultural value orientations, technological preferences, and institutional structures.

The same claims could be made on a global scale. As discussed in Chapter 6, liberal and coordinated welfare and production regimes hold different ecological potentials which, in turn, are exploited in regime-specific ways. From this perspective, international treaties on environmental protection should acknowledge that different regulatory approaches to ecological issues exist and – for that reason – leave sufficient room for habitat-specific adaptation measures rather than prescribing one and the same remedy for all.

Bibliography

Bettencourt, L. M., Lobo, J., Helbing, D., Kühnert, C., & West, B. (2007). Growth, innovation, scaling, and the pace of life in cities. *Proceedings of the National Academy of Sciences*, 104(17), 7301–7306.

Krugman, P. (2009). The increasing returns revolution in trade and geography. *The American Economic Review*, 99(3), 561–571.

Norgaard, R. B. (1994). *Development betrayed: The end of progress and a coevolutionary revisioning of the future* (1st ed.). London, New York: Routledge.

(1997). A coevolutionary environmental sociology. In M. Redclift, & G. Woodgate (Eds.), *The International Handbook of Environmental Sociology* (pp. 158–168). Cheltenham: Edward Elgar.

Sieferle, R. P. et al. (Ed.) (2006). *Das Ende der Fläche: Zum gesellschaftlichen Stoffwechsel der Industrialisierung*. Cologne: Böhlau.

Appendix

Table A.1 Carbon emissions: how much do social and physical concentration actually explain?

1998	Mobility			Housing			Mobility and housing		
	1	2	3	1	2	3	1	2	3
Size of municipal district (reference: <5,000)									
5,000–20,000	150.91***	-4.78	12.71	-44.09	-85.12	-124.82+	-138.73	-213.99*	-233.68*
	(38.51)	(36.28)	(34.988)	(80.75)	(77.10)	(68.31)	(111.09)	(107.23)	(93.53)
20,000–100,000	15.27	-17.23	-13.89	-554.35***	-238.37***	-131.80*	-627.45***	-335.11***	-171.94*
	(35.43)	(33.20)	(31.93)	(68.34)	(66.49)	(58.37)	(97.61)	(94.27)	(82.20)
100,000–500,000	-170.05***	-87.02**	-86.74**	-708.56***	-200.97**	-4.62	-881.25***	-393.82***	-68.36
	(36.45)	(33.13)	(32.05)	(71.39)	(69.17)	(60.77)	(99.28)	(95.58)	(82.35)
≥ 500,000	-288.14***	-181.73***	-185.68***	-654.83***	-247.92**	-46.35	-1078.64***	-688.95***	-269.08**
	(40.44)	(38.96)	(37.44)	(87.50)	(83.64)	(72.80)	(116.07)	(111.12)	(96.19)
Household income per capita		0.55***	0.34***	1.02***	0.97***	0.24***	1.69***	1.65***	0.56***
		(0.02)	(0.02)	(.06)	(0.05)	(0.05)	(0.07)	(0.07)	(0.07)
German nationality (1 = not German)		56.61+	-79.55**	145.95*	319.50***	-139.67**	243.70**	449.43***	-249.19***
		(30.51)	(30.37)	(57.97)	(54.05)	(46.09)	(79.70)	(75.58)	(70.95)
Age		-9.99***	-14.71***	18.26***	18.80***	2.00	5.17*	5.14**	-12.74***
		(0.75)	(0.81)	(1.64)	(1.52)	(1.31)	(2.09)	(1.97)	(1.81)
Cars per household		697.34***	892.48***				458.31***	399.31***	903.70***
		(16.54)	(20.26)				(43.55)	(42.16)	(47.06)

Education (reference: secondary I "Hauptschule")								
Secondary II ("Realschule")	27.84	20.28	−158.88**	−85.93+	−57.59	−124.31	−35.14	−55.99
	(29.38)	(28.42)	(55.34)	(51.32)	(44.39)	(76.53)	(73.52)	(64.31)
High school ("Abitur")	49.96*	73.06*	−200.11**	−226.18***	−158.29*	−119.21	−118.72	−105.57
	(37.37)	(35.45)	(74.29)	(67.86)	(621.88)	(98.01)	(91.98)	(81.83)
Household size (ln)		−889.00***			−931.89***			−1672.52***
		(45.48)			(97.85)			(131.75)
Living space per capita					44.87***			52.70***
					(3.29)			(4.24)
House: year of construction (reference: houses built before 1918)								
1918–1948			−84.59	−133.63+	43.62	−182.42+	−226.54*	−57.98
			(81.48)	(77.13)	(66.69)	(110.69)	(107.22)	(90.95)
1949–1971			−207.47**	−315.73***	−80.67	−264.15**	−378.92***	−107.76
			(72.21)	(69.56)	(60.94)	(100.22)	(99.78)	(87.32)
1972–1980			−131.78	−282.97***	−38.11	−183.26	−323.47**	−51.02
			(84.15)	(82.05)	(70.97)	(115.15)	(115.82)	(99.59)
1981–1990			−729.69***	−574.02***	−154.81+	−886.58***	−680.88***	−195.22+
			(94.41)	(95.13)	(85.43)	(127.67)	(129.87)	(114.20)
1991–2000			−834.07***	−599.44***	−263.31***	−744.56***	−495.88***	−89.37
			(82.82)	(77.75)	(69.30)	(124.53)	(120.48)	(105.30)
Heating (reference: district heating)								
Oil				1760.33***	1522.06***		1893.01***	1524.06***
				(65.62)	(59.95)		(87.12)	(78.24)
Gas				136.54**	−49.75		163.27*	−121.08+
				(48.23)	(46.43)		(68.30)	(64.14)
Electricity				288.73	238.55		413.20+	317.07
				(183.21)	(173.09)		(231.46)	(214.53)
Coal/timber				461.94***	367.57***		413.19***	188.61+
				(94.36)	(85.23)		(121.16)	(109.58)

Table A.1 (cont.)

1998	Mobility			Housing			Mobility and housing		
	1	2	3	1	2	3	1	2	3
Building type (reference: detached single- and two-family houses)									
Single- and two-family row houses						18.64			62.09
						(61.25)			(92.42)
Apartment buildings (3–4 flats)						−28.13			−24.53
						(80.07)			(101.65)
Apartment buildings (5–8 flats)						−196.50**			−234.27*
						(72.61)			(93.84)
Apartment buildings (≥9 flats)						−184.20*			−273.69**
						(75.93)			(96.20)
Constant	1373.39***	−2.61	1274.37***	1413.55***	454.02***	1934.38***	1789.16***	838.63***	2995.54***
	(23.78)	(52.98)	(91.15)	(116.16)	(122.90)	(256.56)	(180.34)	(186.66)	(319.10)
N	14304	12270	12270	6878	6878	6809	6445	6445	6384
R^2	0.008	0.325	0.371	0.234	0.350	0.533	0.302	0.372	0.541
Adjusted R^2	0.008	0.325	0.371	0.232	0.349	0.531	0.300	0.371	0.540

Note: simple linear regressions with carbon emissions (kg CO_2 per capita and year) as dependent variable. Robust standard errors in parentheses. * $p \leq 0.05$, ** $p \leq 0.01$, *** $p \leq 0.001$.

Source: compiled by the author. GSOEP 1998.

Table A.2 Munich: correspondences among dwelling technologies, cultural preferences, and modes of social reproduction

	Building type				
	Detached house	*Row or semi-detached house*	*Detached two- or three-family house*	*Apartment building (up to five stories)*	*Apartment building (six stories or more)*
Number of observations	42	58	51	813	79
Dwelling technologies					
Heating source[1]					
Firewood	0.19	0.05	0.04	0.03	0.01
Oil	0.17	0.16	0.10	0.08	0.04
Gas	0.52	0.64	0.45	0.48	0.20
District	0.02	0	0.10	0.14	0.33
Household size	2.60	2.36	2.00	1.93	1.76
Living space (sq m/capita)	60.98	56.82	66.65	47.86	47.96
Cars (per capita)	0.69	0.71	0.93	0.72	0.82
Cultural preferences					
Percentage of property owners	0.79	0.61	0.43	0.21	0.12
Percentage of tenants	0.21	0.39	0.57	0.79	0.88
Political preference[2]					
Center-right	0.33	0.40	0.46	0.25	0.32
Center-left	0.18	0.14	0.13	0.20	0.18
Grüne	0.30	0.29	0.20	0.33	0.20
Marital status[3]					
Single	0.10	0.05	0.14	0.24	0.22
Unmarried with partner	0.10	0.16	0.14	0.28	0.17
Married	0.71	0.58	0.46	0.31	0.30
Divorced	0	0.09	0.14	0.07	0.14
Church attendance[4]					
Weekly	0.10	0.14	0.06	0.04	0.05
Monthly	0.10	0.12	0.14	0.07	0.04
Seldom	0.71	0.55	0.51	0.65	0.67
Never	0.10	0.19	0.28	0.22	0.21
Modes of social reproduction					
Age					
Men	68.44	62.16	62.46	51.85	55.15
Women	56.82	55.77	56.85	49.23	53.47
Retired					
Men	0.65	0.52	0.42	0.27	0.38
Women	0.29	0.42	0.28	0.22	0.29
Income[5,8]	1.46	2.30	1.93	2.07	1.64
Level of education (middle school or below)[6,8]					
Men	0.13	0.27	0.21	0.18	0.20
Women	0.08	0.26	0.50	0.24	0.42

	Building type				
	Detached house	Row or semi-detached house	Detached two- or three-family house	Apartment building (up to five stories)	Apartment building (six stories or more)
Level of education (university degree)[6,8]					
Men	0.88	0.73	0.79	0.79	0.73
Women	0.92	0.73	0.50	0.74	0.58
Type of employment[7,8]					
Men, full-time	0.75	0.86	0.93	0.84	0.87
Men, part-time	0	0.07	0	0.04	0
Women, full-time	0.25	0.47	0.50	0.63	0.77
Women, part-time	0.33	0.40	0.28	0.18	0.09
Children[8]	0.93	1.68	0.83	1.16	1.50
Care of old or ill relatives[9]					
Yes	0.76	0.77	0.75	0.65	0.50
No	0.07	0.16	0.06	0.18	0.29
CO_2 emissions per capita (t/year)[10]					
Mobility	3.45	3.21	3.71	3.42	3.79
Housing	2.99	3.38	2.81	1.81	1.84
Mobility and housing	3.98	4.63	4.32	2.96	2.85
Consumption	1.40	1.81	1.74	1.85	1.54

Source: compiled by the author. Munich & Bolzano dataset.

[1] Relative frequencies of heating source in percent (within the respective building type). Because not all heating sources are reported here, column percentages do not exactly add up to 100%.

[2] Relative frequencies of political preferences (within the respective building type). Because not all parties are reported here, column percentages do not add up to 100%. Grüne: Greens.

[3] Relative frequencies of marital status (within the respective building type). Because not all categories are reported here, column percentages do not add up to 100%.

[4] Relative frequencies of church attendance.

[5] Expressed in 1000 €/month per capita.

[6] Relative frequencies of men and women of the respective educational level (within the respective building type).

[7] Relative frequencies of the respective type of employment (within the respective building type).

[8] Here, only gainfully employed people are included – otherwise, the large number of pensioners would bias the proportions between male and female full- and part-time employment and corresponding numbers of children.

[9] Here, people were asked whether they could rely on their families in case of care dependency, for example, due to illness or age-related frailty (relative frequencies). Because the category "Maybe" is not reported here, column percentages do not add up to 100%.

[10] As a result of different numbers of observations, the subtotals of CO_2 emissions resulting from mobility and housing do not add up. Different numbers of observations result from the fact that not everybody uses a car.

Table A.3 Bolzano: correspondences among dwelling technologies, cultural preferences, and modes of social reproduction

	Building type				
	Detached house	Row or semi-detached house	Detached two- or three-family house	Apartment building (up to five stories)	Apartment building (six stories or more)
Number of observations	35	31	88	718	204
Dwelling technologies					
Heating source[1]					
Firewood	0.46	0.23	0.20	0.02	0.02
Oil	0.20	0.16	0.15	0.10	0.11
Gas	0.31	0.65	0.60	0.68	0.69
District	0	0	0.01	0.03	0.04
Household size	3.60	3.06	2.76	2.45	2.45
Living space (sq m/capita)	58.45	46.23	51.23	43.31	41.92
Cars (per capita)	0.92	0.66	0.72	0.74	0.74
Cultural preferences					
Percentage of property owners	0.97	0.85	0.84	0.80	0.79
Percentage of tenants	0.03	0.14	0.15	0.20	0.21
Political preference[2]					
Center-right	0.45	0.35	0.35	0.16	0.08
Center-left	0.07	0.15	0.12	0.17	0.32
Grüne	0.14	0.15	0.10	0.13	0.09
Marital status[3]					
Single	0.03	0.03	0.08	0.10	0.07
Unmarried with partner	0.03	0.07	0.10	0.13	0.13
Married	0.77	0.87	0.67	0.58	0.59
Divorced	0	0	0.03	0.05	0.04
Church attendance[4]					
Weekly	0.29	0.19	0.31	0.21	0.23
Monthly	0.34	0.19	0.18	0.13	0.17
Seldom	0.26	0.58	0.45	0.56	0.53
Never	0.06	0	0.05	0.08	0.06
Modes of social reproduction					
Age					
Men	58.60	55.10	55.71	58.40	62.42
Women	49.57	51.60	55.30	53.94	56.33
Retired					
Men	0.22	0.30	0.36	0.45	0.60
Women	0.08	0.27	0.25	0.31	0.35
Income[5,8]	1.52	1.01	2.12	2.59	2.91
Level of education (middle school or below)[6,8]					
Men	0.50	0.29	0.16	0.17	0.24
Women	0.17	0.25	0.42	0.22	0.17

Table A.3 (cont.)

	Building type				
	Detached house	*Row or semi-detached house*	*Detached two- or three-family house*	*Apartment building (up to five stories)*	*Apartment building (six stories or more)*
Level of education (university degree)[6,8]					
Men	0.50	0.71	0.84	0.82	0.76
Women	0.83	0.75	0.58	0.77	0.81
Type of employment[7,8]					
Men full-time	0.93	0.79	0.88	0.92	0.95
Men part-time	0	0.07	0.06	0.03	0.05
Women full-time	0.42	0.63	0.38	0.54	0.47
Women part-time	0.50	0.13	0.46	0.30	0.40
Children[8]	1.80	1.73	1.61	1.14	1.20
Care of old or ill relatives[9]					
Yes	0.88	0.93	0.84	0.84	0.84
No	0.03	0	0.05	0.07	0.06
CO_2 emissions per capita (t/year)[10]					
Mobility	7.37	4.03	3.63	3.01	2.85
Housing	1.84	1.99	2.52	2.24	2.14
Mobility & housing	2.85	3.15	3.36	2.94	2.94
Consumption	2.42	.97	2.44	2.43	2.58

Source: compiled by the author. Munich & Bolzano dataset.

[1] Relative frequencies of heating source in percent (within the respective building type). Because not all heating sources are reported here, column percentages do not exactly add up to 100%.

[2] Relative frequencies of political preferences (within the respective building type). Because not all parties are reported here, column percentages do not add up to 100%. Grüne: Greens.

[3] Relative frequencies of marital status (within the respective building type). Because not all categories are reported here, column percentages do not add up to 100%.

[4] Relative frequencies of church attendance.

[5] Expressed in 1000 €/month per capita.

[6] Relative frequencies of men and women of the respective educational level (within the respective building type).

[7] Relative frequencies of the respective type of employment (within the respective building type).

[8] Here, only gainfully employed persons are included – otherwise, the large number of pensioners would bias the proportions between male and female full- and part-time employment and corresponding numbers of children.

[9] Here, people were asked whether they could rely on their families in case of care dependency, for example, due to illness or age-related frailty (relative frequencies). Because the category "Maybe" is not reported here, column percentages do not add up to 100%.

[10] As a result of different numbers of observations, the subtotals of CO_2 emissions stemming from mobility and housing do not add up. Different numbers of observations result from the fact that not everybody uses a car.

Index